Engineering GNVQ: Advanced

620 TOO

Engineering GNVQ: Advanced

Mike Tooley
Director of Learning Technology
Brooklands College of Further and Higher Education

Newnes

OXFORD   AUCKLAND   BOSTON   JOHANNESBURG   MELBOURNE   NEW DELHI

Newnes
An imprint of Butterworth-Heinemann
Linacre House, Jordan Hill, Oxford OX2 8DP
225 Wildwood Avenue, Woburn, MA 01801-2041
A division of Reed Educational and Professional Publishing Ltd

 A member of the Reed Elsevier plc group

First published 2000

© Reed Educational and Professional Publishing 2000

British Library Cataloguing in Publication Data
A catalogue record for this book is available from the British Library.

ISBN 0 7506 4755 8

Printed and bound in Great Britain

Edexcel recommends this book as suitable for its
Advanced GNVQ Engineering/Vocational A-Level programme.

FOR EVERY TITLE THAT WE PUBLISH, BUTTERWORTH-HEINEMANN
WILL PAY FOR BTCV TO PLANT AND CARE FOR A TREE.

Contents

Preface

Welcome to the challenging and exciting world of engineering! This book is designed to help get you through the core units of the Advanced General National Vocational Qualification (GNVQ) in Engineering. It contains all of the material that makes the essential underpinning knowledge required of a student who wishes to pursue a career in any branch of engineering.

This book was originally written by a team of Further and Higher Education Lecturers. They each brought with them their own specialist knowledge coupled with a wealth of practical teaching experience. This new edition has been designed to cover the core units of the revised and updated GNVQ Engineering programme. Throughout we have adopted a common format and approach with numerous examples, problems and activities.

About GNVQ

General National Vocational Qualifications (GNVQ) are available in schools and colleges throughout England, Wales and Northern Ireland. The main aim of these 'vocational A-levels' is that of raising the status of vocational education in the UK within a new system of high quality vocational qualifications which can be taken as an alternative to the well-established General Certificate of Education (GCSE) and General Certificate of Educational Advanced Level (GCE A-level) qualifications.

With the advent of the latest Government initiative, *Curriculum 2000*, GNVQ awards have taken on an important new role in a highly flexible curriculum for students aged 16 to 19. It is now possible to mix vocational and non-vocational studies and many students will, for the first time, be able to try their hand at engineering whilst studying for more conventional A-level qualifications.

As well as acquiring the basic skills and body of knowledge that underpin a vocational area, all GNVQ students have to achieve a number of *key skills*. The attainment of both vocational *and* key skills provides a foundation from which students can progress either to further and higher education, or into employment with further training appropriate to the job concerned.

GNVQs, like NVQs, are unit-based qualifications. Each is made up of a number of units which can be assessed separately, and this allows credit accumulation throughout a course. A certificate can be obtained for each unit if necessary and credit transfer on some units can be made between qualifications. To make this possible, assessment is based on the unit rather than the qualification.

How to use this book

This book covers the six mandatory units that make up the GNVQ Advanced Engineering programme. One chapter is devoted to each unit. Each chapter contains text, worked examples, 'test your knowledge' questions, activities, and multi-choice practice questions.

The worked examples will not only show you how to solve simple problems but they will also help put the subject matter into context with typical illustrative examples.

The 'test your knowledge' questions are interspersed with the text throughout the book. These questions allow you to check your understanding of the preceding text. They also provide you with an opportunity to reflect on what you have learned and consolidate this in manageable chunks.

Most 'test you knowledge' questions can be answered in only a few minutes and the necessary information, formulae, etc. can be gleaned from the surrounding text. Activities, on the other hand, require a significantly greater amount of time to complete. Furthermore, they often require additional library or resource area research coupled with access to computing and other information technology resources.

Activities make excellent vehicles for gathering the necessary evidence to demonstrate that you are competent in core skills.

Finally, here are a few general points worth noting:

- Allow regular time for reading – get into the habit of setting aside an hour, or two, at the weekend to take a second look at the topics that you have covered during the week.

- Make notes and file these away neatly for future reference – lists of facts, definitions and formulae are particularly useful for revision!

- Look out for the inter-relationship between subjects and units – you will find many ideas and a number of themes that crop up in different places and in different units. These can often help to reinforce your understanding.

- Don't expect to find all subjects and topics within the course equally interesting. There may be parts that, for a whole variety of reasons, don't immediately fire your enthusiasm. There is nothing unusual in this, however do remember that something that may not appear particularly useful now may become crucial at some point in the future!

- However difficult things seem to get – don't be tempted to give up! Engineering is not, in itself, a difficult subject, rather it is a subject that *demands* logical thinking and an approach in which each new concept builds upon those that have gone before.

- Finally, don't be afraid to put your new ideas into practice. Engineering is about *doing* – get out there and *do* it!

Good luck with your GNVQ Engineering studies!

Mike Tooley

Unit 1 Engineering in business and the environment

Summary

All engineering companies must operate as commercial enterprises in order to survive. In this unit you will look at how engineering companies are structured, and how they combine commercial and engineering functions to meet their objectives. This will be underpinned by an understanding of how engineering activities as a whole contribute to the economy. You will also learn about the importance of effective communication, and about how companies make important financial decisions. In addition, you will learn about environmental legislation and how it affects the engineering industry, and about how engineering companies design their operations to minimize negative environmental effects of their activities.

Economic contribution of engineering activities

Engineering affects all industries that make or use engineered products. Engineering activities therefore span a huge number of industries and sectors, including the following key engineering industries:

- electrical and electronic
- mechanical
- aerospace
- telecommunications
- motor vehicle
- maintenance
- chemical
- marine.

You must be able to recognize which industry a particular company belongs to by considering either what they produce or what their main functions are. In order to gain an awareness of the economic importance of engineering activities, you must also be able to find out how individual industries contribute to the economy and employment in the UK. We shall begin by explaining some basic economic concepts.

Economic concepts

National product

This is an important term that warrants some explanation. To help us in our explanation we will show how the process of manufacturing a wooden chair contributes to the national product.

Figure 1.1 shows the complete production process for the manufacture and sale of a simple wooden chair. The process is broken down into four stages with the first being the supply of the raw timber, the tree. This supposes that the tree-grower pays his workers their wages and retains a business profit by selling the tree to the timber merchant for a total price of £30. The timber merchant adds to the £30 his costs of sawing the tree into pieces, storing the cut timber and his profit. The added value in this case is £20 making the subsequent sale price to the chair manufacturer £50.

The chair manufacturer adds a further £30 of production costs and profit making the wholesale value of the chair £80. The furniture shop retailing the chair to the public adds his shop and distribution costs and his profit to the £80 making a high street selling price for the finished chair of £100.

The £100 price of the chair is the money value of the output or product of the chair making process. The process has contributed £100 to the national product. Note that this £100 can be calculated either by taking the final selling price or by adding together the

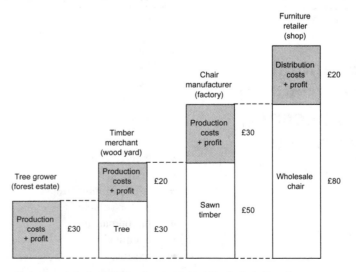

The gross product value of £100 can be calculated by taking the final selling price or by adding together the added values (shaded) contributed by each of the four stages of production.

Figure 1.1 *Diagram showing the national product arising from the production of wooden chairs*

shaded areas of *added values* contributed by each of the four different stages of chair production.

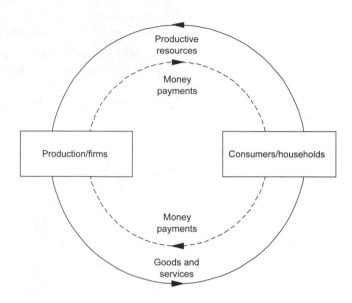

Figure 1.2 *Flow of goods, services and productive resources in an economic system*

Figure 1.2 shows how the production of goods and services and their consumption can be regarded as forming a closed circle. The firms make the goods and provide the services consumed by the people living in the households. The people living in the households are the same ones that own and work in the production and service firms. The demand for the goods and services is originated by the households which is satisfied by the production firms that pay the wages of the householders. Hence, the flow of goods and services and production resources are in one direction and is balanced by the flow of money in the other.

Yet another diagram encountered in written articles concerning the economy – the management of the material resources of a community or country – is shown in Figure 1.3. In many respects Figure 1.3 is no more than a combination of Figures 1.1 and 1.2. The diagram again relates to the production the same wooden chairs as discussed earlier. The four major production functions are shown but this time with the added value element at each stage being sub-divided into wages and profit. The wages are paid to the stage production workers and the profit to its shareholders. Of course, all of these people are themselves buyers of the wooden chairs and hence the arrows in Figure 1.3 showing the wages and profits of each stage of production being passed to the users of the wooden chairs.

The horizontal arrows linking the production stages represent the cash payments made for the materials supplied. You will notice how the flow of total output money from each production stage is exactly matched by the money it receives from the previous production stage. For example, in the case of the final production stage, the buyer of a wooden chair pays the furniture retailer £100. Of this amount, the furniture retailer pays the chair manufacturer £80, for the wholesale chair, and pays his shop workers and shareholders the remaining £20 in wages and profit dividend respectively. So, the national product for the whole engineering function of wooden chair production, £100, can calculated either by:

• the final expenditure (the retail price of a chair) method, or
• the sum of the added values method.

Figure 1.3 allows us to see both of these methods.

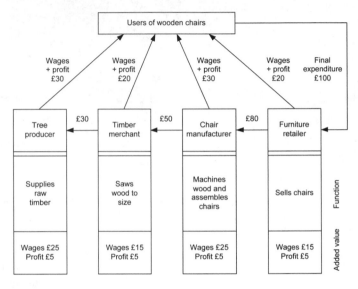

Note: The users of wooden chairs are the same people that own, and provide the labour force for, the different stages of production.

Figure 1.3 *An alternative way of representing the engineering function of producing wooden chairs*

Gross national product (GNP)

The total output of the UK can be measured by adding together the value of all goods and services produced by the UK. This figure is called the gross national product. The word 'gross' implies that no deduction has been made for the loss in value of the country's capital equipment, which helps to make the national product, caused by normal wear and tear.

The word 'national' in this context does not mean that the GNP is the total output produced within the borders of the UK. You see, the gross product of the UK includes some output taking place and produced by resources within the UK but owned by people from other countries. Therefore, this particular element of output cannot be regarded as part of our national income. At the same time, some sources of output are located in other countries but owned by UK citizens. Therefore the GNP is defined as the value of the total output of all resources owned by citizens of the UK wherever the resources themselves may be situated.

Gross domestic product (GDP)

If we measure the value of the total output actually produced within the borders of the UK, it is called the gross domestic product. So, the GDP is defined as the value of the output of all resources situated within the UK wherever the owners of the resources happen to live. In many ways the GDP is a more important measure than the GNP.

The government periodically issues statistical tables showing the performance of the different sectors of the economy. You will find this in the 'Blue Book', issued by the Central Statistics Office, entitled *United Kingdom National Accounts*. Further information is also available from the Government Statistical Office via the World Wide Web at http://www.statistics.gov.uk/stats.

Activity 1.1

The UK's GDP for the 11 year period from 1983 to 1993 are shown in the Table 1.1 below. Plot these figures on a line graph. Also determine:

(a) the percentage increase in GDP over the 10 year period from 1984 to 1993
(b) the average annual percentage increase in GDP over the 10 year period from 1984 to 1993
(c) the year in which the actual increase in GDP was least
(d) the year in which the actual increase in GDP was greatest.

Present your results in a brief written report. Hint: You may find a spreadsheet useful for determining annual changes and also for plotting graphs.

Table 1.1 *UK GDP for the 11 years from 1983 to 1993*

Year	GDP (£ million)
1983	261,225
1984	280,653
1985	307,902
1986	328,272
1987	360,675
1988	401,428
1989	441,759
1990	478,886
1991	495,500
1992	516,027
1993	546,120

Gross domestic product per head

This is a measure of *productivity*. It relates output to the number of people employed producing that output. The formula used is:

Output per head (or *per capita*) = output produced divided by the number of people producing it.

Put another way, we can write:

Output per head = output / employment.

Often we are concerned not with absolute figures for output and employment but more with trends. For this reason, both the output and employment figures are usually quoted as *index* figures (i.e. they are stated relative to a base figure of 100 for a particular year). Table 1.2 shows how the UK output, employment, and output per head has changed over the eight year period from 1990 to 1997.

Table 1.2 *UK output, employment and productivity for the period 1990 to 1997*

Year	Output	Employment	Output per head
1990	100.0	100.0	100.0
1991	100.5	100.1	100.4
1992	100.6	100.0	100.6
1993	100.5	99.5	101.0
1994	99.9	99.4	100.5
1995	90.5	90.0	100.6
1996	89.5	88.9	100.6
1997	89.5	88.8	100.8

Activity 1.2

The regional variation of GDP per capita in the UK for 1991 and 1996 is shown in Table 1.3 below. Plot these figures on two bar charts in descending order of per capita GDP in each case. Use the data to determine:

(a) the best and worst performing regions for each of the two years
(b) the region that has shown the greatest increase in per capita GDP over the three year period
(c) the region that has shown the greatest decline in per capita GDP over the three year period

Present your results in a brief written report. Hint: You may find a spreadsheet useful for determining the three year changes and also for plotting graphs.

Table 1.3 *Regional variation of GDP per capita in the UK for 1993 and 1996*

Region	Per capita GDP (UK = 100)	
	1993	1996
North East	85.8	84.3
North West	91.1	90.9
East Midlands	96.7	94.3
West Midlands	92.8	93.5
London	126.6	123.3
South West	95.2	94.7
South East	111.0	111.4
Northern Ireland	81.2	81.2
Scotland	98.6	99.1
Wales	82.7	83.1

Direct employment

This is the term used to describe the mode of employment of people that are actually working to produce a product. For example, the people working in a factory making furniture could all be regarded as direct labour being directly employed on the manufacture of the furniture.

Indirect employment

Following on from the previous example, the people concerned with the transportation of the production materials to the furniture factory and the finished furniture to the shops which sell the furniture, are regarded as indirect labour and are in indirect employment.

Note however, that people may be directly employed by their own trade but indirectly by another. For example, the people felling the trees and producing the raw timber used to make the furniture are directly employed by the timber trade but are indirectly employed by the furniture making trade.

Exports

Exports are goods and services that we sell to other countries. Visible exports are the hard goods which are physically transported abroad and for which we receive payment. Invisible exports are those services that we provide for foreigners and for which they pay. Invisible exports include the payments we receive for insurance and financial services, technical or military training and the like. Tourism is also an important part of our invisible export trade. A German tourist in London will spend money that he has earned in Germany. This provides a useful income to the UK no different from, say, visibly exporting and selling the German a bottle of Scotch whisky in Berlin. Note that when we take our holidays abroad, the goods and services we buy and consume abroad are in fact imports to the UK.

Imports

Imports flow in the opposite direction to that of exports. Imports are the goods and services that we purchase from abroad.

Balance of payments

If we are to remain financially sound, we must ensure that our income exceeds our expenditure. This applies both to our personal domestic lives and to the nation as a whole. On a national basis, the difference between the total income we earn from abroad for our exports is balanced against the total payment we must make abroad for our imports. We compare or *balance* the two figures and hopefully the export figure is at least as large as and preferably larger than the import figure.

Very often the UK balance of visible exports and imports shows an excess of import value over exports. However, we have a relatively healthy sale of our services abroad and our *invisibles* usually gives us a favourable trading balance. The government lets us know what our overseas trading position is by issuing monthly balance of payment figures.

Engineering sectors

The engineering sectors and typical engineered products with which you need to be familiar are as follows:

Chemical
Fertilizers, pharmaceuticals, plastics, petrol, etc. Companies in this field include Fisons, Glaxo, ICI and BP.

Mechanical
Bearings, agricultural machinery, gas turbines, machine tools and the like from companies such as RHP, GKN and Rolls-Royce.

Electrical and electronic
Electric generators and motors, consumer electronic equipment (radio, TV, audio and video) power cables, computers, etc. produced by companies such as GEC, BICC and ICL.

Civil
Concrete bridges and flyovers, docks, factories, power stations, dams, etc, from companies like Bovis, Wimpey and Balfour-Beatty.

Aerospace
Civil and military aircraft, satellites, space vehicles, missiles, etc. from companies such as British Aerospace, Westland and Rolls-Royce.

Telecommunications
Telephone and radio communication, data communications equipment, etc. from companies like Nokia, GEC, Plessey and BT.

Motor vehicles

Cars, commercial vehicles (lorries and vans), motorcycles, tractors and specialized vehicles from companies such as Rover, Vauxhall UK and McLaren.

Activity 1.3

Marine engineering is another important engineering sector. Prepare a brief word processed report suitable for a foreign investor who is considering investing in the UK marine engineering sector. You need not include detailed financial information but you should include a description of the sector that identifies *at least four* marine engineering activities in the UK. You should also include contact and other information for *at least three* UK engineering companies that are active in this field.

Activity 1.4

A group of Estate Agents has asked you to carry out some research for them based on your home town. They are particularly interested in the size and scope of local engineering firms with a view to future developments.

(a) Identify the three largest engineering firms in the area of your home town.
(b) Find out who owns these firms and whether they are part of a larger group of companies.
(c) Allocate each firm to the appropriate engineering sector.
(d) Ascertain the annual sales turnover (its output in money terms) of each firm.
(e) Draw a histogram showing the individual engineering sector turnover in your area.

Present your findings in the form of a brief word processed report and include relevant diagrams and tables.

Activity 1.5

Visit your local museum and obtain information concerning the engineering companies that were active in your area (city, town or county) in 1900 and in 1950. Allocate each of these companies to a particular engineering sector and compare your findings with the present day situation. Identify any new industries and explain why they have developed in your area. Present your findings in the form of a class presentation with appropriate visual aids.

The different economies

You need to be familiar with the effects of engineering at four different economic levels: local, regional, national and European. In each of these economies the engineering activities appear to be spread unevenly. This is the case whether we examine a local town or rural area, a larger region of possibly two or three counties, the whole of the UK on a national basis or the European Union.

Local economy

For the first half of the twentieth century, engineering was generally located within cities. Since then there has been a tendency for any new engineering enterprise to be located in an industrial estate on the periphery of a town rather in the town itself. Usually this is because:

- the town centre is already too congested to allow for additional new industry;
- of the advantages of being located in a ready, purpose-built industrial accommodation on a site having good road links with the national motorway network;
- engineering activities which may involve noise and other pollutants are best kept away from the commercial and domestic centres of towns.

In general, the engineering industries that remained in the city centres have slowly become outdated and, in many cases, have closed down. The impact of this migration from the city centres to the suburbs has been to leave derelict buildings, unemployment and social deprivation for the city residents. For the outer suburbs receiving the new engineering industries, the impact has not always been positive. The decentralization of engineering from the city centres has contributed to urban sprawl, and this has led to conflict for land on the city's boundary between engineering, farming and recreation. Also, it has tended merely to move the problem of engineering pollution from the city centre to its suburbs.

Regional economy

The regional economy comprises many local economies but the change in the engineering pattern is much the same. While there is still a great deal of engineering activity to be found in and around many large cities and built-up areas, there is a definite migratory move towards the small town and rural areas. This trend is to be found in most economically developed countries and has been a consistent feature of the last 25 years.

National economy

At the national level the uneven spread of engineering is between the different regions. The processes that caused this variation are historic. Very often they are directly connected to the availability of natural resources. For example, in the nineteenth century, regions rich in coal were favoured with engineering expansion because of the local availability of coal to fire boilers to drive the steam engines that powered the factories. The technical skills acquired by the workers in the coal bearing regions were the same skills required for other industries and enterprises and cumulative expansion took place. This expansion, and the highly paid work it created, attracted labour from the less industrialized regions so exacerbating the regional disparities.

However, over the last 50 years there has been a shift of engineering away from the old industrial regions such as the North-East and Midlands of England and parts of Scotland to more convenient locations such as the Thames Valley along the M4 motorway and along the M11 motorway north of London. The reasons for the regional shift are many and varied and include such factors as:

- because of its cost and pollution causing record, coal is no longer a popular fuel;
- with natural gas and electrical power being available almost anywhere in the country, new engineering activities can be located in regions having pleasant natural and social environments;
- because of the ubiquitous motor car, good roads and frequent air services, commuter and business communications to most regions are no longer a major problem;
- the availability of a pool of technologically skilled labour in places where high-technology companies are clustered together.

European Union

Within the EU, engineering activities have the usual varied pattern. The favoured countries are those which were the first to industrialize in the eighteenth and nineteenth centuries. Britain, Germany, France and Italy are predominant in Europe with the main concentration lying within a rough triangle formed by London, Hamburg and Milan. Ireland, Spain, Southern Italy and Greece lie outside this triangle and tend to be less industrialized.

The past 30 years has seen a shift in some of the major engineering activities which used to be concentrated in Europe, North America and Japan. In particular, much of the electronics and printing industries have migrated to the *Pacific rim* countries such as Hong Kong, Singapore, Taiwan, Thailand and more recently into Indonesia. The main reason for this shift is the low labour costs to be found in the Far East.

Another prime example of the shift of engineering activities out of Europe is that of shipbuilding. Britain's contribution in particular has fallen and is now virtually nonexistent except for the manufacture of oil platforms and ships for the Royal Navy.

However, the traffic in engineering activities has not been all negative. The Japanese, wanting to sell their motor cars in Europe, have established several engineering production plants in Britain. The firms of Toyota (Deeside and Burnaston), Nissan (Sunderland) and Honda (Swindon) are three good examples. All occupy rural sites and have access to skilled and well-educated workforces. Road communications are good and, in the case of Nissan, the site is in an assisted area where substantial Government grants are available.

Activity 1.6

A company based in the Far East has asked you to carry out some research in order to help them investigate some investment opportunities in Europe. They have asked you to produce a broad comparison of the performance of the various countries that comprise the European Union (EU).

(a) Draw a map of Europe (Hint: Photocopy or scan this from a suitable atlas).
(b) Mark in the countries that constitute the EU.
(c) Mark in the capital cities of each EU country.
(d) Ascertain the GDP for each EU country then list the countries in descending order of GDP. What is the position of the UK?

Produce your findings in the form of a set of overhead projector transparencies and handouts to be used at a board meeting.

Test your knowledge 1.4

Why do Japanese firms need to manufacture their cars in the UK as opposed to Japan?

Commercial and engineering functions in engineering companies

You need to be able to identify both commercial and engineering functions within an engineering company. Commercial functions can include sales, marketing, distribution, commissioning, finance and purchasing. Engineering functions can include research and development, design, manufacturing, product development, quality, and planning. Product support is another, extremely important, area of engineering, and you should be aware that non-engineering companies often require input from engineering functions, such as for maintenance of services and equipment. You must also be able to recognize the main responsibilities attached to key job roles within both commercial and engineering functions.

Management activities

Planning

Planning is absolutely fundamental to the correct functioning of an engineering company. If no planning is done then activities are almost certainly going to be very ineffective. What is planning? It is the sum of the following activities:

- setting the *goals* for an engineering company
- forecasting the *environment* in which the engineering company will operate
- determining the *means* to achieve goals.

Setting goals or objectives must be the first step. It determines the direction an engineering company is going. It encourages all engineering company members to work toward the same ends, otherwise members are likely to set their own objectives which will conflict with each other. Good objectives make for rational engineering companies that are more co-ordinated and effective. The objectives must therefore be set by the most senior management group in an engineering company so that all of its staff can be given clear direction. If the goals are clearly stated, logical and appropriate for the business then they act both as motivators and yardsticks for measuring success.

Once the engineering company has clear direction the next step is to analyze the environment and forecast its effect on the business. For example an engineering company that makes lawn mowers may set objectives so that, within five years, it will:

- achieve a 30% share of the market
- be acknowledged as the market leader
- be the accepted technological leader
- be highly competitive on price
- operate internationally.

When it forecasts its environment it may conclude that:

- new designs will be marketed by competitors
- new battery technology will become available to support cordless mowers
- there will be a sharp decline in demand for manual lawn mowers.

Its designs are technically very sound but are being threatened by new rotary designs which are proving attractive to customers. It has to decide how to deal with the threat, either to improve its existing design concept so that customers continue to find them attractive or to follow the new trend and produce products based on new design concepts.

Forecasting the environment allows the company to set new objectives and to prepare plans to meet its revised goals. Companies that fail to go through this process will go into decline in the long run because they are ignoring the changing world around them.

Once the goals have been refined and changed in the light of environmental forecasting then plans can be made to achieve the goals. Some plans will not change that much, others will be dramatically affected by the changing environment. For this reason plans can be classified as follows:

- standing plans
- single use plans
- strategic plans
- tactical plans.

Standing plans are those that are used many times, and remain relatively unaffected by environmental change. Examples are employment, financial, operating and marketing policies and procedures. For example hiring new employees involve standard procedures for recruitment and selection. Another example would be the annual routines for establishing budgets.

Single use plans are those that are used once only, such as those for the control of a unique project or specific budgets within an annual budget. (Budgets themselves are single use plans, even though the procedures used for producing them are standing plans.)

Strategic plans are the broad plans related to the whole engineering company and include forecasting future trends and overall plans for the development of the engineering company. They are often in outline only and very often highly subjective, involving judgements made by top managers. For example a plan may be made to build an entirely new factory based on forecasts of demand. This plan is strategic, and if it is wrong and the sales forecasts on which it is based do not materialize, the results for a company could be devastating.

Tactical plans operate within the strategic plan. The new factory has to be brought into commission and production has to be scheduled and controlled. Plans for the latter are tactical, since they focus on how to implement the strategic plan.

Control

The pre-requisite of control is planning. Controlling involves comparing events with plans, correcting deviations and ensuring that the planned events happen. Sometimes deviations are so fundamental that they require a revision to the plan so that later events are controlled against a new plan. For example the original sales forecast may turn out to be too optimistic, and production plans may have to be reduced to bring output into line with what sales are possible.

There are various ways in which control can be exercised. It can be predictive as in the case of a cash-flow forecast. This forecast may indicate a shortfall of cash in August but a surplus in September. The finance manager may need to arrange additional finance with the bank in August and then in September he might deposit surplus funds onto the money market. The point here is that variances are predicted in advance, thereby promoting cash control.

In the case of monthly comparisons between budgeted expenditures and actual expenditures an overspend might be revealed. This triggers action that holds back expenditure until spending comes back into line with budget. This is historical control since action is based on a report of events in the recent past.

Concurrent control is *real time* such as that which might occur in controlling a continuous process or a production line. In this case the system has built in *feedback* which enables it to remain in balance by regulating inputs and outputs. An example of concurrent control would be where a production process requires temperature regulation. The control system is designed to switch off heating when the temperature reaches a threshold or switch on heating when it drops to a minimum level. The "feedback" is information on temperature. The same principle applies in some stock control systems, where stocks are maintained at pre-determined minimum and maximum levels, with supplies being switched on and off to maintain equilibrium.

Leadership and direction

Planning and control activities are the tasks of management. However, they are only achieved through people. People will work effectively if they are led and directed properly. This implies that top managers must be in touch with the business and be visible. They must have a clear vision for the future, reinforced by specific objectives which are communicated to their employees.

This approach to leadership is apparent in some of our best companies as exemplified by Marks & Spencer. Such companies have a clear mission and objectives, and have a visible committed top management. This philosophy permeates the whole engineering company stimulating better performance from all employees.

Motivating good performance from all employees is the responsibility of all managers. What motivates individuals and groups within commercial engineering companies is a complex and important subject, the detail of which is well beyond the scope of this book. However, it is still worth saying that managers must discover what it is that will stimulate employees to work productively.

In general people respond best to 'considerate' styles of management, whereby their personal contributions are fully recognized. It is also true that there has to be an atmosphere of discipline coupled with a work-oriented culture. The task has to be accomplished, and being considerate does not extend to the toleration of slack or sloppy practices and behaviour.

Clear direction, sound and explicit guidelines, well worked out procedures all of which are well communicated, ensure that the engineering company works smoothly.

Allocation and supervision of work

This is the practical implementation of all that we have discussed in this section. An engineering company exists to fulfil the goals of its owners. It has to function in a co-ordinated and rational way. The people who are its members have to work together and understand their specific roles and functions. They need to receive directions and work has to be allocated. There has to be supervision of these activities. An engineering company is analogous to a machine or a living organism. In order to function properly everything has to work together smoothly. The managers have the task of ensuring that this work takes place according to plan and within the engineering company's stated objectives.

Activity 1.7

A case study on the organization of an engineering company

A company makes diesel engines for use in a variety of products and environments. It had one division which shipped *knocked down* (KD) kits to overseas assemblers (these assemblers were foreign subsidiaries of the company). These firms assembled the kits into working engines and sold them on to other manufacturers for incorporation into their products. The kits were put together based on engineering specifications for each type of engine. A pre-production control department was responsible for translating the engineering specifications into packing lists used to prepare and pack the kits for export. Engineering specifications were constantly being updated to improve designs and to correct faults that came to light during service. Changes often had to be carefully planned so that the KD parts shipped were compatible and of the same engineering level.

The KD operation received many complaints about the kits. Some assemblers said they could not assemble the engines because of specification errors or because the wrong parts were shipped. The pre-production control department was very skeptical about these complaints and was slow to respond. A proportion of the complaints were ill-founded so requests for urgent replacements were given a low priority and normal packing operations deadlines were considered more important than sending out urgent miscellaneous orders to enable the overseas assembly locations to complete their work.

The management were under heavy pressure to meet their own packing deadlines which were constantly threatened by supply problems of all kinds and by frequent changes to orders from overseas assemblers. The kits normally took between six weeks and three months to reach their destination by sea-freight, whilst miscellaneous orders had to be air-freighted at high cost.

The problems continued to worsen as complaints from overseas companies grew. These companies tried to anticipate problems by comparing the packing documentation with the manufacturing engineering process sheets and faxing in queries on all discrepancies. This generated so many queries that it was impossible to respond quickly even if pre-production activities wished to. This actually generated even more inertia on their part. In the end Head Office appointed a senior manager as troubleshooter and sent him in to solve the problems.

What is wrong with the organization of this engineering company?
What role should the manufacturing engineering activity play here?
What should be the troubleshooter's first priority?
What do you think the troubleshooter should do in order to arrive at a long term solution?
What steps should the company take to avoid this sort of problem in the future?

Discuss this problem in small groups and prepare a 10 minute presentation to the rest of your class or group giving details of your proposed solution.

Commercial functions

This section examines some of the main functional areas of businesses and shows the interface between them. It discusses the variety of jobs done by engineers in business, and how engineering techniques are used with other techniques in fulfilling the needs of the business.

Finance and accounting

The *Finance Director* will manage the company's cash flow, short term and long term investments and supervise the company's accounting and budgetary system. His or her job is to ensure that the company has sufficient cash to support day-to-day operations. He should also ensure that cash that is not immediately required is made to work for the company. This is done by arranging short term investments on the money markets or by switching funds into building society or bank deposit accounts. He may also be involved with international money flows and the *hedging* of risk against exchange rate losses.

The *Financial Accountant* will be in charge of all recording in the financial accounting system. All business transactions are recorded in a system called *double entry accounts*. These records are called this because every business transaction has a two-fold effect, and to make a complete record this two-fold aspect has to be recorded. For example if £100 was spent to buy raw materials, there has to be a record in the bank account of the money spent and then the existence of materials bought must be recorded in a purchases account. This method of recording enables the business to produce a profit and loss account and a balance sheet, which summarize all the transactions so financial performance can be tracked and reported to shareholders and other interested parties, including the tax authorities.

The *Management Accountant* will administer the budgetary control and costing system. This system enables the business to forward plan its profit and loss and its overall financial position, as well as being able to control and report on costs of operation. Thus a budgeted profit and loss account and balance sheet is produced, called a *Master Budget*. This Master Budget summarizes the budgets for all cost centres or operating divisions. The management accountant monitors actual results against the budget and will use data from the double entry accounts system referred to above. This monitoring enables departmental managers to correct deviations to budget, control and manage the cost of running their departments.

Part of management accounting is the process of investment appraisal to plan the purchase of fixed assets and to ensure the best choices are made for new and replacement equipment. The management accountant will also prepare reports and analyses of various kinds for management.

The *Cashier*, who is likely to report to the financial accountant will deal with all bank and cash transactions. In a large company the number of transactions is considerable, especially those that deal with receipts from customers and payments to suppliers. The work of this section is important because forecasting and monitoring cash flow is vital to the financial well-being of the business.

The *Credit Controller* will be concerned with authorizing new credit customers and controlling the amount of credit granted and reducing or preventing bad debts. A budget will be prepared giving planned levels for outstanding debtors. This figure really represents a short term investment of capital in customers. This investment has to be managed within budgeted limits otherwise the company's finance costs would increase and there would be further risk of non-payment. The Credit Controller would monitor the financial stability of existing customers or vet the standing of new customers. He might do this through a combination of bank references, credit agencies and studying the customers' own published accounts.

Other activities carried out in accounting include payment of wages and salaries, depreciation of fixed assets, maintaining shareholder records, paying shareholder dividends.

A typical engineering company structure for the accounting and finance function is given in Figure 1.4 which shows the departmental divisions indicated above.

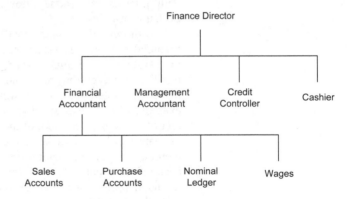

Figure 1.4 *Organization of the finance function*

Marketing, sales and distribution

Marketing

Marketing is said to be the most important function in the business, since if customers cannot be found for the company's products the company will go out of business, regardless of how financially well run or efficient it is.

Although sales is considered separately later, it is really part of the marketing function. Marketing is all about matching company products with customer *needs*. If customer needs are correctly identified and understood, then products can be made which will give the customer as much as possible of what he wants. Companies which view the customer as *sovereign* are those companies that stay in business, because customers will continue to buy products that meet their requirements.

Hence marketing activities are centred around the process of filling customers' known needs, discovering needs the customer does not yet know he or she has and exploiting this by finding out how to improve products so that customers will buy this company's products in preference to other goods. Some of the most important activities are:

- market research
- monitoring trends in technology and customer tastes
- tracking competitors' activities
- promotion activities
- preparing sales forecasts.

Remember that in some businesses the marketing activity is directed at end consumers, members of the public. This has to be done by national forms of advertising, such as TV commercials, direct mail, newspapers or through major retailers selling to the consumer. The methods used may be somewhat different if the customers are other companies. Although the principle of meeting customer needs is the same, the approach taken may be much more technical and may include the services of sales engineers to provide technical back-up and advice. The publicity methods are more likely to be centred around trade fairs, exhibitions, advertisements in the trade press or technical journals, for example. You should note these two distinct marketing approaches are respectively called consumer marketing and industrial marketing.

Sales

The sales department is concerned with advertising and selling goods. It will have procedures for controlling sales and the documentation required. The documents used are the same as for purchasing, described later, except from the supplier's viewpoint rather than from the customer's. It may employ commercial travellers or have a resident sales force. It is involved with many possible ways of publicizing the company's products such as trade fairs, wholesalers' displays, press and TV advertising, special campaigns, promotional videos, etc. It will also be concerned about the quality of goods and services as well as administering warranty and guarantee services, returns and repairs, etc.

Activity 1.8

Interview the person responsible for marketing in your school or college. Prepare a set of questions that will help you to obtain a view as to the importance of the marketing function in your school or college and the activities that are included within it. Obtain a selection of marketing materials and find out how they are used to attract students, parents and employers. Produce a questionnaire and use it to determine the effectiveness of the course leaflet or prospectus that is used to market your own GNVQ course. Identify any weaknesses and suggest ways in which this material can be improved. Prepare a brief word processed report to present your findings.

Sales will maintain contacts with customers that will entail the following customer services:

- technical support
- after-sales service, service engineering
- product information, prices and delivery
- maintaining customer records.

A typical Marketing and Sales organizational structure for a company engaged in industrial marketing is shown in Figure 1.5.

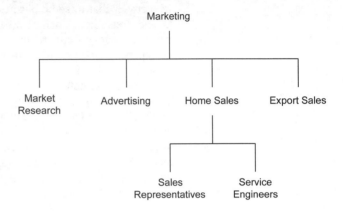

Figure 1.5 *Organization of the marketing and sales functions*

Distribution — consumer markets

For companies operating in the consumer marketing field distribution can be accomplished through a variety of ways. This can include wholesalers, retailers, mail order or direct selling through the company's own retail outlets. Some companies may use all of these methods. We will examine the wholesale and retail systems, as well as the pricing aspects.

Retail outlets

These outlets sell direct to the consumer. Most of these are shops or mail order businesses. Retailers fall into several types, hypermarkets, supermarkets, multiple shops, departmental stores, Cooperative Retail Societies, Independent Retailers, Voluntary Retail Chains, Franchise Outlets, Discount Stores.

The purpose of retailing is to provide for the availability of goods close to where consumers live. Retailers also study consumer preferences and stock goods accordingly. They also keep manufacturers informed of what it is that consumers want so that supply matches demand.

Wholesale outlets

If a retailer requires a large range of goods in relatively small quantities it is not very convenient to buy direct from manufacturers. Think of the number of different manufacturers that a small independent grocer would have to deal with if he did deal direct with each manufacturer. Hence the continuing need for wholesalers, who stock goods from many manufacturers and can supply smaller quantities to retailers.

The wholesaler is a middle man and it is said that his presence puts up prices. This is not necessarily the case, since manufacturers can sell to him in bulk quantities and save on transport and administration costs. In effect wholesalers operate as intermediate storage depots for retailers and therefore provide a useful service. They can usually provide retailers with credit terms of trading, often enabling small businesses to sell before they have to pay for goods, or at least to reduce the impact of the cost of carrying a large range of stock items.

They can also act as a buffer to smooth out demand for manufacture. If demand is seasonal they can buy regularly through the year, thus making it easy for manufacturers to make goods in economic runs and then store stock to meet heavy demand, but which does not place excessive loads on the manufacturer's capacity.

Wholesalers as such have been in decline in recent times, thus many manufacturers have started to deal direct, especially with large retailers, such as the supermarkets. However, they have had to take over the functions of storage, transport and dealing directly with retailers.

Mail order

Companies may decide to deal directly with the public through mail order, thus by-passing the wholesaler and retailer. Mail order depends on a good postal service or the existence of transport operators who can provide a similar service. It has the advantage of being nation-wide or even international, thus extending the potential market enormously.

In some cases there are very large mail order retailers who buy from manufacturers and sell on to consumers. These companies sometimes operate normal retail outlets as well.

Price structure

It is common for distributor prices to be expressed at a percentage discount from the price to be paid by the consumer. Thus manufacturers give discounts to wholesalers as an incentive to stock their goods and to provide a profit margin for him. A similar system will be used by the wholesaler when dealing with the retailer. However, if a price to the final consumer is not envisaged or fixed, then the situation is less clear and each party must charge a price which his particular market will stand and depending on what quantities he needs to sell and what his actual costs are going to be.

In addition to wholesale and trade discounts, quantity discounts may be offered to encourage distributors to buy in large quantities which may be more economical to supply and deliver, since there are economies of scale to be had, such as lower manufacturing, administrative and transport costs.

Further incentives may be offered in giving cash discounts. Cash discounts reward immediate or early payment. This may enable the manufacturer or wholesaler to reduce his need for working capital and reduce credit collection and control costs.

Discount structures are therefore used for many purposes. First, to increase sales, second to influence the pattern of sales and third to reduce the costs of production and distribution. Sometimes discounts can produce more sales but have very little effect on profits since higher volumes and lower costs may not compensate for lower margins.

You should be aware that price is influenced by many factors:

(a) actual cost of manufacture
(b) what the market will stand
(c) what others are charging for similar products
(d) consumers' perceptions of quality and value.

The interaction of supply and demand is complex and outside the scope of this course. However, if supply exceeds demand, in general this exerts a downward pressure on prices, as manufacturers and distributors seek to sell goods they have made or bought. The costs of storage and distribution may so high as to force sale at prices which might be below average cost. This is especially true of perishable goods and foodstuffs.

Alternatively, if demand exceeds supply that tends to bid up the price as consumers search for supplies. In some cases the increase in price then serves to limit demand since some potential buyers drop out when the price goes too high, this then acts to dampen demand again and tends to bring equilibrium between supply and demand.

As you can appreciate this can get very complicated when manufacturers, wholesalers and retailers start to offer the different discounts to try to influence events in their favour.

Sometimes the competition is so cut throat that the only winner is the consumer. In some cases the weaker players go under, leaving the more efficient firms to operate in a less hostile environment.

Sometimes the bigger stronger firms deliberately cut prices so low as to force others out of business and then exploit the consumer when they can dominate the market. However, this can go in reverse again if then prices go too high and this attracts new players into the market who will then increase supply which will then produce a downward pressure again on prices, and so it goes on.

Distribution — industrial markets

For industrial market distribution the situation is more variable. Frequently the seller will have his own fleet of vehicles and may have warehousing facilities or geographically dispersed depots. An example might be a company which manufactures components for the motor industry. It may manufacture in one or more locations and have storage depots located near its customers. It may also deliver products direct to the motor manufacturer's plant either using its own transport or by using an independent haulier.

If the company makes products for international markets it may have to prepare and package products for sea or air freight. This could include using haulage contractors who will deliver direct into Europe using roll-on roll-off ferries.

Price structure
This is also very variable, but is usually based on negotiation between the seller and buyer. It may be done through a process of enquiry and quotation or may simply be based on price lists and discounts separately negotiated.

Research and development (R&D)

New product design and development is often a crucial factor in the survival of a company. In an industry that is fast changing, firms must continually revise their design and range of products. This is necessary because of the relentless progress of technology as well as the actions of competitors and the changing preferences of customers.

A good example of this situation is the motor industry. The British motor industry has gone through turbulent times, caused by its relative inefficiency compared with Japan and Germany and also because the quality of its products was below that of its competitors. This difficult position was then made worse by having products that lacked some of the innovative features of the competition.

Strategies for product development

There are three basic ways of approaching design and development:

- driven by marketing
- driven by technology
- co-ordinated approach.

A system driven by marketing is one that puts the customer needs first, and only produces goods which are known to sell. Market research is carried out which establishes what is needed. If the development is technology driven then it is a matter of selling what it is possible to make. The product range is developed so that production processes are as efficient as possible and the products are technically superior, hence possessing a natural advantage in the market place. Marketing's job is therefore to create the market and sell the product.

Both approaches have their merits, but each of them omit important aspects, hence the idea that a co-ordinated approach would be better. With this approach the needs of the market are considered at the same time as the needs of the production operation and of design and development. In many businesses this inter-functional system works best, since the functions of R&D, production, marketing, purchasing, quality control and material control are all taken into account. However, its success depends on how well the interface between these functions is managed and integrated. Sometimes committees are used, as are matrix structures or task forces (the latter being set up especially to see in new product developments). In some parts of the motor industry a function called *programme timing* co-ordinates the activities of the major functions by agreeing and setting target dates and events using network planning techniques.

The development process

The basic process is outlined as follows:

- idea generation
- selection of suitable products
- preliminary design

Test your knowledge 1.6

1 What is the most important function of marketing?
2 What is the purpose of market research?
3 Why do some companies use wholesalers rather than sell direct?
4 Name two ways of determining the price of a product.
5 What does a sales engineer do?

- prototype construction
- testing
- final design.

This is a complex process and involves co-operative work between the design and development engineers, marketing specialists, production engineers and skilled craft engineers to name some of the major players.

Ideas can come from the identification of new customer needs, the invention of new materials or the successful modification of existing products. Selection from new ideas will be based on factors like:

- market potential
- financial feasibility
- operations compatibility.

This means screening out ideas that have little marketability, are too expensive to make at a profit and which do not fit easily along-side current production processes.

After this, preliminary designs will be made within which trade-offs between cost, quality and functionality will be made. This can involve the processes of *Value Analysis* and *Value Engineering*. These processes look at both the product and the methods of production with a view to maintaining good product performance and durability whilst achieving low cost.

Prototypes are then produced, possibly by hand and certainly not by using mass production methods. This is followed by rigorous testing to verify the marketing and technical performance characteristics required. Sometimes this process will involve test marketing to check customer acceptance of the new product.

Final design will include the modifications made to the design as a result of prototype testing. The full specification and drawings will be prepared so that production can be planned and started.

Activity 1.9

A local engineering company has asked you to assist with the development of a new product. This product is to be aimed at the DIY market and is to consist of a combination steel rule, drill gauge and spirit level.

(a) Suggest ways of determining the market potential and financial feasibility of this product.
(b) Identify the materials and processes that could be used to manufacture the product – are these materials and processes available in your school or college? Could you produce a prototype with the resources that are available to you?
(c) Explain why it would be necessary to test market this product.

Present your findings in a brief word processed report.

Manufacturing operations

The production or manufacturing operation is at the heart of the business. It translates the designs for products, which are based on market analysis, into the goods wanted by customers.

The production process can be seen within a framework of five main areas which we will now discuss.

Process and facilities management

Decisions have to be made in relation to location of the factory and the design and layout of production facilities. The design of production processes is interactive with product design, requiring close co-operation with R&D and marketing functions.

Selecting the process of production is important and is strategic in nature. This means that it has a wide impact on the operation of the entire business. Decisions in this area bind the company to particular kinds of equipment and labour force because the large capital investments that have to be made limit future options. For example a motor manufacturer has to commit very large expenditures to lay down plant for production lines to mass produce cars.

Once in production, the company is committed to the technology and the capacity created for a long time into the future. There are three basic methods for production processes:

*   line flow
*   intermittent flow
*   project.

Line flow is the type of system used in the motor industry for assembly lines for cars. It also includes continuous type production of the kind that exists in the chemicals and food industries. Both kinds of line flow are characterized by linear sequences of operations and continuous flows and tend to be highly automated and highly standardized.

Intermittent flow is the typical batch production or job shop, which uses general purpose equipment and highly skilled labour. This system is more flexible than line flow, but is much less efficient than line flow. It is most appropriate when a company is producing small numbers of non-standard products, perhaps to a customer's specification.

Finally *project-based production* is used for unique products which may be produced one at a time. Strictly speaking there is not a flow of products, but instead there is sequence of operations on the product which have to be planned and controlled. This system of production is used for prototype production in R&D and is used in some engineering companies who produce major machine tool equipment for other companies to use in their factories.

Capacity planning

Once facilities for production have been put in place the next step is to decide how to flex the capacity to meet predicted demand. Production managers will use a variety of ways to achieve this from maintaining excess capacity to making customers queue or wait for goods to having stocks to deal with excess demand. The process is complex and may require the use of forecasting techniques, together with careful planning.

Scheduling activities are different for each process method and require the use of a variety of techniques. The objectives of good scheduling are:

- meeting customer delivery dates
- correct loading of facilities
- planning starting times
- ensuring jobs are completed on time.

Inventory control

With any manufacturing facility good inventory control is an absolute essential. It is estimated that it costs up to 25% of the cost value of stock items per year to maintain an item in stock. Proper control systems have to be used to ensure that there is sufficient stock for production while at the same time ensuring that too much stock is not held. If stock levels are high there are costs associated with damage, breakage, pilferage and storage which can be avoided.

Work force management

This is related to the need to have a labour force trained to use the facilities installed. The important aspects here are:

- work and method study
- work measurement
- job design
- health and safety.

The production manager has to establish standards of performance for work so that the capacity of the factory can be determined and so that the labour costs of products can be calculated. Work study, method study and work measurement activities enable this to be done, as well as helping to promote efficient and safe methods of working.

The design of jobs is important in respect of worker health as well as effective work. Good job design can also make the work more interesting and improves employee job satisfaction, which in turn can improve productivity.

Quality control

Quality is a key objective for most engineering companies. It is especially important to the production function that is actually manufacturing the product for the customer.

What is meant by the word quality? It is generally defined as *fitness for purpose*. In effect this means meeting the identified needs of customers. Thus it is really the customer that determines whether or not a company has produced a quality product, since it is the customer who judges value received and registers satisfaction or dissatisfaction.

This does bring problems for manufacturers since customer perceptions of quality vary, some customers will like a product more than other customers will. Hence a manufacturer has to use some more objective criteria for assessing fitness for purpose. It has been suggested that this must include:

- design quality
- conformance quality
- reliability
- service.

Design quality is the primary responsibility of R&D and Marketing. It relates to the development of a specification for the product that meets identified needs. *Conformance quality* means producing a product that conforms to the design specification. A product that conforms is a quality product, even if the design itself is for a cheap product. That seems contradictory, but consider the following example. A design is drawn up for a budget camera, which is made from inexpensive materials and has limited capability. If the manufacture conforms to the specification then the product is of high quality, even though the design is of low quality compared with other more up-market cameras.

Reliability includes things like continuity of use measured by things like *mean time between failure* (MTBF). Thus a product will operate for a specified time, on average, before it fails. It should also be maintainable when it does fail, either because it can easily and cheaply be replaced or because repair is fast and easy. *Service* relates to after sales service, guarantees and warranties.

Quality control is therefore concerned with administering all of these aspects. In the UK there are general standards for *quality systems*, the most relevant one here is BS 5750 and the international counterpart, ISO 9000. The activities of quality control include the following:

- inspection, testing and checking of incoming materials and components
- inspection, testing and checking of the company's own products
- administering any supplier quality assurance systems
- dealing with complaints and warranty failures
- building quality into the manufacturing process.

Whilst many of these activities are performed in order to monitor quality after the event, others may be carried out to prevent problems before they occur and some may be carried out to determine causes of failure that relate to design rather than manufacturing faults.

Activity 1.10

Use your library or other information sources to find out how an engineering company can become registered under ISO 9000. Present your findings in the form of a brief class presentation with appropriate visual aids.

Purchasing and supply

In large businesses purchasing is done by professional buyers, and is therefore a centralized activity. When a company has a large purchasing budget this makes economic sense, since the large purchasing power gives advantages in negotiating for keen prices, better delivery times or increased quality. In small businesses the purchasing function is not centralized, usually because the operation is not large enough to support the employment of specialists. However, the basic principles of purchasing are the same, whatever the structure of the engineering company.

Main functions

The main functions are:

- researching sources of supply
- making enquiries and receiving quotations
- negotiating terms and delivery times
- placing contracts and orders
- expediting delivery
- monitoring quality and delivery performance.

The basic documents used are as follows:

- requisitions from departments to buyer
- enquiry forms or letters to suppliers
- quotation document in reply to enquiry
- order or contract to buy
- advice note – sent in advance of goods.
- invoice – bill for goods sent to the buyer
- debit note – additional/further charge to invoice
- credit note – reduction in charge to invoice
- consignment note – accompanies goods for international haulage, containing full details of goods, consignee, consignor, carrier and other details
- delivery note – to accompany goods.

Purchasing procedures

Purchasing procedures involve raising a requisition to buy. This may then require obtaining quotations or estimates in order to choose the best supplier. Orders are sent to the chosen supplier. Goods are dispatched by the supplier, together with a delivery or advice note. When goods are received they are checked to see that the details on the delivery note agree with the actual goods received and that the goods have in fact been ordered. Goods not ordered may be refused. Accepted deliveries are signed for on the supplier's copy and given back to the driver. A goods received note is raised and sent to the purchasing department, so that the accounting function can be given confirmation of delivery before making payment against receipt of invoice.

These procedures are for the purpose of making sure that the goods are delivered on time, in the correct quantities, of the correct

---

Test your knowledge 1.8

1 Explain the purpose of the following production functions:
(a) manufacturing/production engineering
(b) capacity planning
(c) inventory control
(d) quality control.
2 State and briefly describe the three methods of process design.
3 Name three techniques used in work-force management.

specifications and of the desired quality. Only if all is well is payment authorized.

A typical large, centralized purchasing engineering company would be as shown in the engineering company chart in Figure 1.6. It is common for there to be a purchasing manager and buyers who will specialize in particular types of commodity. Sometimes the products being purchased require that the buyer is technically qualified.

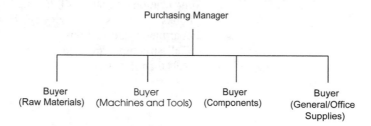

Figure 1.6 *A centralized purchasing function*

Organizational structures

Understanding how a particular engineering company operates is made easier by referring to its organizational structure. There are various ways of describing these structures and we will consider these one at a time. Note that the descriptions are not necessarily mutually exclusive, thus one description may also overlap with other descriptions.

Flat structures

Flat structures have few levels from top to bottom. Figure 1.7 shows a flat organizational structure in which five managers report directly to a *Board of Directors*. The five managers are described as *first line managers* because they directly supervise the activities of people who actually do the job. Such people are often described as *operatives*.

Figure 1.7 *A flat structure*

The 1990s in the UK saw a significant move towards having flatter organizational structures. The process has been described as *delayering* and it is regarded as an organization structure that permits better communications both up and down the organizational levels. It is also seen as more cost effective and responsive in dealing with demands placed on modern businesses.

Note that flat organizations are also hierarchical. A hierarchy is an organization with grades or classes ranked one above another. A flat organization also meets that criterion.

Tall structures

These are opposite to flat structures and contain many layers. Figure 1.8 shows an example. Tall structures usually exist only in large organizations because of the necessity of dividing the tasks into chunks of work that can be handled by individual managers, departments and sections.

Tall structures are also hierarchies, only they contain many more levels than flat structures.

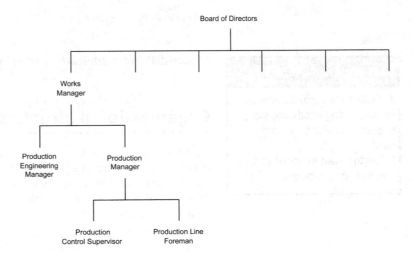

Figure 1.8 *A heirarchical structure*

These are structures which have many or few layers showing rank order from top to bottom. They show a chain of command, with the most senior posts at the top and the most junior posts at the bottom. Plain hierarchies are the most common representations of organizations, as some aspect of hierarchy exists in all organizations.

The *organization chart* is a useful way of representing the overall structure, but it tells only part of the story. You should be aware that other documents are needed to fully understand how the organization works, as we observed earlier.

Hierarchical organizations can take many forms. We have already examined flat and tall structures. There are several other forms with which you should be familiar. One of the most common is the functional design. Figure 1.9 shows a functional organization.

The main functions of a commercial business are marketing, finance, purchasing and supply, manufacturing, research and development and personnel. Notice how each functional manager reports to the managing director who co-ordinates their activities. There are a number of advantages in functional structures:

- specialists can be grouped together
- it appears logical and easy to understand

- co-ordination is achieved via operating procedures
- suits stable environments and few product lines
- works well in small to medium size businesses.

Figure 1.9 *A functional structure*

As a business grows the functional structure becomes less and less useful. This is because there are many more products and these may be manufactured in separate divisions of a company, especially if economies of scale are introduced into the manufacturing process.

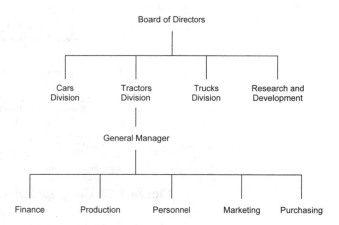

Figure 1.10 *A federal or divisional structure*

Figure 1.10 shows an organization based on major product lines and is really a federal structure which still has functional activities, but at a lower level in the organization, except for the research and development (R&D) function which is centralized. The managers of these operating divisions will control most of the functions required to run the business.

In many conglomerate businesses this federal arrangement is achieved by having a holding company, which may be a *public limited company* (plc), which wholly owns a number of subsidiary companies, which are in effect divisions of the main business. Figure 1.11 shows an example.

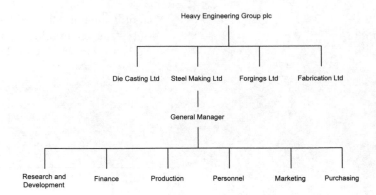

Figure 1.11 _A conglomerate organization_

An alternative to the product based divisional design is one based on geographical divisions. Figure 1.12 shows a geographical design which still has many of the functions located at a head office, but which has branches dispersed around the country. These branches or divisions handle sales and manufacture, but are supported by head office for the other functions.

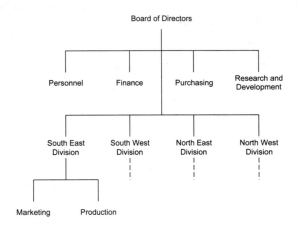

Figure 1.12 _Geographical divisions_

Matrix structures

A matrix structure as its name implies has a two-way flow of authority. Departmental or functional authority flows vertically and project management authority flows horizontally. Such a structure is shown in Figure 1.13. This depicts a company that designs and makes machines and tools to customer specifications.

Some businesses operate as a series of projects. This is common, for example, in construction and in some types of engineering company, especially those that design one-off products or who design manufacturing equipment used by other manufacturers. As the chart shows, there are the familiar functional divisions, and members of functional departments still report to their functional line manager. However, the project manager has the job of co-ordinating the work of functional specialists to ensure that the project is completed on time, to specification and within cost limits.

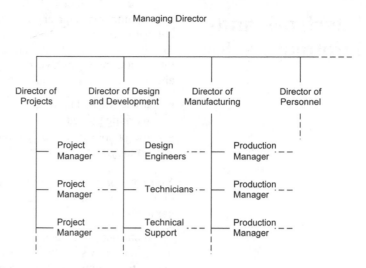

Figure 1.13 *A matrix structure*

In the main, project managers do not have direct line authority, but have to influence and persuade others to achieve *targets*. They may have formal authority over project budgets and can set time schedules and decide what work is to be done, but little else. They also work as an interface with clients and subcontractors and their influence is often critical to the success of the project. Although they do not have formal authority over individual staff or their line managers, they never-the-less operate with the full support of senior managers. This means that functional specialists are obliged to provide the fullest co-operation and help, otherwise they become answerable for failure to their own senior line managers.

The matrix system works very well in project based industries, and that is why the design is used. It still retains many of the ingredients of other structures, and still has substantial hierarchical elements.

Hybrid structures

The structures discussed above are just examples of the main design principles for organizations. There are numerous variations and rarely do we find 'pure' forms of organization structure. We need to remember that organizations are created to serve the goals of their owners and that the precise structure will be designed to meet the needs of the business.

Interfaces and information flow

The commercial and engineering functions within an engineering company, as well as individual teams within each, need to work together in order to function effectively and efficiently. This interaction might involve sharing information and making decisions about:

- processes and systems
- working procedures
- the people involved, including customers, suppliers and other employees.

You must be able to identify and describe the working relationships or interfaces that exist between departments within an engineering company.

You also need to be able to recognize types of information and the data that flows between the various departments within an engineering company. These can include:

(a) documents such as:

- design specifications
- purchase orders
- invoices
- production schedules
- quotations.

(b) information about:

- stock levels
- work in progress
- resource utilization sales.

Note that all of the above may exist either as *hard copy* or may be in electronic form (such as a word processed document, a hypertext document, or a spreadsheet file).

Interfaces

Finance and accounting

The finance and accounting function interfaces with all other functions within an engineering company. Its recording and monitoring activities are central to, and have a major impact on, the whole business. In conjunction with the manufacturing function, *sales forecasts* will be used to prepare *factory schedules*. Production may be sent to a warehouse and then put into the delivery and distribution system. From there the sales force will ensure that customers receive their orders when required. Alternatively delivery of specific customer orders may be made directly to customers.

Marketing

Marketing will supply information on prices. Prices may be determined primarily by the market rather than the cost of manufacture. Finance may provide cost information, but marketing may make the final pricing decisions.

Marketing will also identify customer needs and will liaise with product development activities on possible new products or modifications to existing products. R&D will initiate design studies and prototypes for new products and may supply some items for market testing. Engineers involved with design and development will be given information on customer needs and preferences and will be expected to produce designs which meet those requirements.

There will be a need to communicate details of the costs of new products or redesigned products. The processes required to produce new components or whole new products will also require costing. R&D may specify the manufacturing process, but manufacturing engineering departments located at the production facility will implement them, and may also share in the costing process.

Manufacturing and production

New products will have different characteristics, and perhaps be made from different materials from previous products with similar functionality. This will require liaison between design engineers and manufacturing engineering on methods for production and in deciding what manufacturing equipment and machine tools are required. Detailed process sheets may be required which show how products are to be assembled or made.

Whilst the particular methods of production are the province of production management, the designer has to be aware of the implications for his design of different methods of manufacture, whether this be batch production, assembly lines or one-off projects. Detailed specifications of the new and changed product will be communicated and there may be liaison on temporary and permanent deviations to original specifications in order to facilitate production.

When quality problems appear and are related to faulty design there will be liaison on ways in which design modifications can be phased into production as soon as possible.

There will be proposals for the replacement of machines and equipment used for manufacturing and production. This function may require quite sophisticated techniques for what is called *investment appraisal* so that the company can choose the best methods of manufacture from several alternatives.

Also important is the control of raw materials and component stocks, especially the levels of *work-in-process*. Finance will want to restrict stock levels to reduce the amount of capital tied up in stocks, whilst the production manager will be concerned with having sufficient stock to maintain production, but avoiding congestion of factory floor space.

Budgetary control of production cost centres will involve regular contact and advice from the Finance function. Matters of interest will be costs of production, wastage rates, labour costs, obsolescent stock, pilferage, etc.

Purchasing

Specifications and drawings will be sent to the buyer for new products or machines for purchase. Problems of design and delivery will be discussed, modifications to designs will be sent to suppliers through the buying department.

New product launches will be co-ordinated with R&D, the supplier, and of course, the manufacturing department. The buyer would be involved with supplies of new raw materials, new designs for components and will negotiate costs of tooling and long-term contracts.

Liaison with factory management on initial supplies for new materials or components. Assistance may given by the buyer to deal with quality and inspection problems and in dealing with return and replacement of defective materials.

Sometimes buyers may be an interface with production and R&D in dealing with temporary or permanent deviations from the original engineering specifications. Chasing deliveries and ensuring supplies for factory use may be a major daily routine for some buying departments.

Systems for quality control may include some form of supplier quality assurance. The buying department will represent company interests to suppliers and may only use suppliers who have passed the company's quality assurance standards. The buyer will be involved with searches for new suppliers who can meet existing and new quality requirements.

Stock control

Stock control systems used within a manufacturing company will affect the way purchasing is done. Economic order quantities may be established which the buyer has to take into account when arranging supplies. Deliveries may have to be phased according to minimum and maximum and re-order stock levels.

The buyer will need a clear understanding of the importance of deliveries which enable the company to control its inventory costs, while at the same time ensuring a reliable supply of materials and components for production.

The company may operate a *just-in-time* system (JIT). JIT originated in Japan and is a way of delivering supplies at the point in time they are required by production. JIT avoids the costs of holding buffer stocks of raw materials and components.

JIT works well when suppliers are dependable and when transport systems are good. The buyer will liaise with the factory on the establishment and operation of the JIT for given products.

There will also be the routine matters of passing invoices for payment of goods or dealing with returns for credit so that the accounts department can pay for goods received. Materials purchasing will be subject to budgetary constraints like most other company activities. The purchasing department will be involved, either directly or indirectly in budgets for inventory levels, and in setting up minimum, maximum and re-order levels for stocks.

Monthly monitoring of inventory levels will be done by the accounting function and purchasing activities may be responsible for ensuring that stock of components and raw materials stay within agreed levels.

Test your knowledge 1.12

With the aid of a labelled diagram, identify the links that exist between the R&D function and the marketing, manufacturing, and purchasing departments in a typical engineering company. For each link, give examples of the type of information that is transferred between the departments.

Financial decisions

All engineering activities are bound by financial decisions – for example a company will need to decide whether to make a component or to buy one from another company, or whether to invest in a new production facility. They will need to put a clear case for a new product or service at the financial planning stage. Companies use financial information and data to help make these decisions. You need to be familiar with the sorts of data and information used to make decisions about:

- whether a company should 'make or buy', e.g. components, assemblies or services for a product
- the production volume needed to generate a required profit (using break-even analysis)
- investment and operational costs.

Operational factors

The usual techniques for costing, budgeting, inventory control, investment appraisal, make or buy decisions and forecasting are examined and compared using examples drawn from a variety of situations. Cost control is concerned with collecting operating cost data and then monitoring and controlling these costs. Budgeting is the process of forecasting the financial position and providing a plan against which to monitor profitability. Inventory control is the establishment of economic levels of stock whilst maintaining production. Investment appraisal enables managers to choose the best projects for investment using discounted cash-flow techniques. Other techniques are examined that enable choices to be made to make or buy and to forecast sales and production.

Cost control

Cost accounting is part of the management accounting function, indeed, without a system for cost accounting effective management accounting could not exist. Management accounting exists to provide information for the internal control and management of a business. Thus cost accounting is necessary for a company to be able to identify responsibility for costs and to exercise control over actual costs compared with planned expenditure.

This is a very important function in respect of R&D as well as manufacturing activities. Engineers are subject to the discipline of budgetary control as are other specialists and it is essential that costs can be monitored and controlled so that engineering projects, for example, meet target costs and profits.

The elements of cost

A manufacturing example

The elements of cost in a typical manufacturing company can be analysed as follows:

Direct costs (for materials, labour and expenses) are costs which can

Example 1.1 Cost analysis

Cost analysis	£
Direct materials	1,200
Direct labour	800
Direct expenses	200
Prime cost	2,200
Factory indirect expenses	1,100
Production cost	3,300
Administration expenses	1,300
Selling and distribution expenses	2,100
Finance expenses	900
Total cost	7,600

be identified with a particular product or cost unit, and as such would not have been incurred had that product or cost unit not existed.

These costs are quite easy to identify. For example, a car headlamp assembly will be based on an engineering bill of materials, specifying components and quantities required. Similarly the standard times for assembly will be known, either by work measurement or from actual data, so the direct labour cost can be calculated. If the assembly has to be delivered to the customer then the direct expense of the specific delivery can easily be obtained.

However, separately identified *direct expenses* are relatively rare. This is because costs for deliveries, for example, are often included in the buying price for materials or the selling price of finished goods. But something like royalties payable to a patent or copyright owner are common in some industries, such as in publishing. Prime cost is therefore the total of these three elements.

Factory *indirect expenses* (another term used is *manufacturing overhead*) are all the other expenses concerned with the manufacturing process which cannot be traced, or are not actually worth tracing, to the product or cost unit. These costs may include power usage, supervision, plant maintenance, non-production consumable materials, etc. Together with prime cost, indirect expenses make up the total works or production cost of manufacture.

The remaining overheads are for the administration, selling and distribution and finance costs. When added to production cost the whole or total cost of operations is obtained.

Although the example is that of manufacture, these elements of cost are universally applicable to all industries and commercial operations.

Marginal costing

The concept of marginal cost is an important one for engineers and managers. It is an alternative way of looking at costs and provides insights into the way costs behave. In particular it enables the engineer or manager to observe the interaction between costs, volumes and profit. Some costs vary with the level of activity (volume) others do not, thus total costs of products will vary depending on the volume produced. This effect is significant and has an important bearing on decisions such as make or buy, or whether or not to accept lower prices for part of a factory's output.

Accountants define marginal cost as:

The amount by which aggregate costs change if the volume of output is increased or decreased by one unit.

Figure 1.14 shows the behaviour of costs based on the marginal cost concept. The vertical axis of Figure 1.14 shows aggregate costs and revenues in £. The horizontal axis shows the units of output. The fixed costs are shown as a straight line across all levels of output. Fixed costs are costs which are unaffected by activity, at least in the short term. Figure 1.14 assumes that such costs do not vary for the whole range of output possible.

Fixed costs include things like:

- insurance premiums
- business rates
- subscriptions
- audit fees
- rental charges
- fixed elements of power and telephone charges.

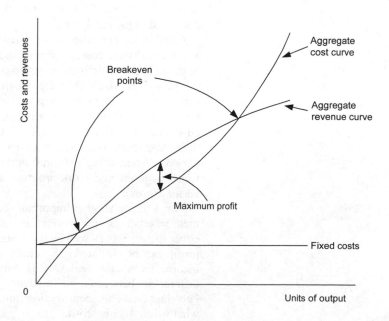

Figure 1.14 *Marginal cost behaviour*

The *aggregate cost curve* starts from the *fixed cost line*. This curve increases steadily with increases in output and for a time can be virtually a straight line. Its slope represents the incremental increase in variable costs as output increases. Variable cost is defined as cost that is incurred to actually make the product, such as direct materials and direct labour. It therefore varies with the level of activity, more product, more materials, etc. However, as volume increases beyond a certain point variable costs increase more steeply, since an extension in capacity is likely to cost more (an example is overtime premiums paid to operatives). Thus the aggregate cost curve becomes steeper. The aggregate revenue curve, which starts from the origin, also increases as output is sold to customers. At a certain point it crosses the aggregate cost line. Here the total cost is equal to total revenue, thus enough units have been produced to recover all fixed and variable costs. Beyond this point revenue is greater than cost so that a *profit* is earned. However, at a certain point sales become more difficult to achieve as the market becomes saturated so that revenue per unit declines. One reason for this would be price discounts to encourage customers to buy more. Economics teaches us that if supply outstrips demand prices will decline and that is what could be happening here.

The slope of the revenue line becomes less steep and eventually crosses the aggregate cost line for a second time. At that point a new breakeven point is reached where revenues and costs are again equal. The point of maximum profit is the point at which the revenue line is above and farthest away from the aggregate cost line. Attempting to push activity beyond this point is self-defeating, since total profit is lower – the business is working harder for fewer gains.

This concept of *marginal costs* forms the basis of an alternative to absorption costing discussed earlier. This new philosophy of costing provides a model of costing which is extremely useful for management purposes. It is also related to other new concepts which help us to more fully understand product costs, and these are discussed immediately below.

Fixed costs are costs which arise out of management decisions and are really the costs of providing productive capacity for a period of time. They are defined as period costs. They have to be charged in the period in which they are incurred, because by definition they cannot be related to any past or future period. *Absorption costing* breaks this principle since products are deemed to absorb period costs and such finished goods are valued at this total cost. Hence these absorbed costs are, in effect, carried forward to succeeding periods. When using marginal costs this does not happen, because finished goods and work-progress are valued at marginal cost (or *variable cost*).

This leads to another important concept. If period costs have to be met, whether or not production takes place, then revenues must cover these costs as well as the marginal (or variable) costs before a profit can be declared. Similarly when sales revenue occurs the income from each unit sold is reduced by its marginal cost. The difference between the revenue and marginal cost is called *contribution*. This contribution initially pays the fixed costs and what is left over is profit.

Marginal costing takes a different view of the world and is also much more sensitive to changes in stock levels and profitability.

Breakeven analysis

Breakeven analysis is an application of the ideas behind marginal costing. The technique is limited but helpful in getting an understanding of cost behaviour and perceiving the interaction between profit, volumes and costs. Because it is frequently presented in graphical form it is used to communicate such information to managers and engineers.

All costs must be covered before a business makes a profit. When contributions from products equal the total costs then it is said that the *breakeven point* has been achieved. This breakeven point can be obtained from a formula as follows:

$$BE\ point = \frac{FC}{SP - VC}$$

Where *FC* is the total fixed cost, *SP* is the selling price per unit and *VC* is the variable cost per unit. Using the selling price the breakeven point in units can be converted into sales value or turnover. Example 1.2 shows how this works.

Example 1.2   Breakeven point

	£
Fixed cost	6,000
Selling price per unit	20
Variable cost per unit	10

$$BE = \frac{6000}{20 - 10} = 600\ units\ = £12,000$$

This can be represented by a breakeven chart as shown in Figure 1.15.

The chart in Figure 1.15 is not unlike Figure 1.14 except that it represents the data using straight lines. The assumption made is that in the period represented in the graph the prices and costs remain the same. Note the shaded areas for *profit and loss*. It is possible to see the degree of profit and loss at different levels of output. You can see also how the composition of cost changes with different levels of output. At low levels fixed costs are a greater proportion and at high levels they are a smaller proportion. Thus the *total cost* per unit decreases as production increases. It is not a static value, thus you can only define total cost for products at given levels of output. This, of course, explains why absorption costing can be so misleading.

However, you should note that breakeven analysis has limitations since it makes the assumption that costs and prices remain unchanged. In reality this is not the case. Also if the chart has been drawn based on a portfolio of products then any change in their mix will alter the relationships shown. It also assumes that everything that is made is also sold, a very big assumption. But in spite of its limitations breakeven analysis does provide some useful insights into how costs behave and does help to make cost information meaningful.

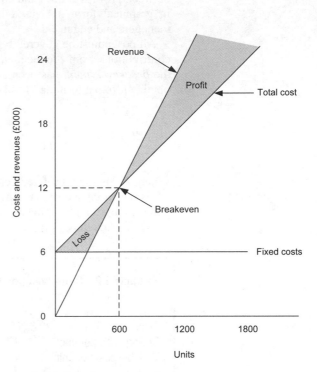

Test your knowledge 1.13

Explain the following terms:
(a) fixed cost
(b) marginal cost
(c) breakeven.

Figure 1.15 *Breakeven chart*

Test your knowledge 1.14

Explain TWO limitations of breakeven analysis.

Activity 1.12

Use the following cost data and construct a breakeven chart using suitable graph paper. Identify the breakeven point in units and sales turnover. Verify the result of your chart by using the breakeven formula.

Fixed costs: £10,000
Variable cost per unit: £5
Sales price per unit: £10

Total sales (maximum capacity) is 5,000 units

Depreciation of fixed assets

Depreciation is the estimate of the cost of a fixed asset consumed during its useful life. If a company buys a car, to be used by a sales representative, for £15,000 it has to charge the cost in some reasonable way to the profit being earned. This process is essential, otherwise the whole cost of running the business cannot be obtained, and profit figures would be overstated. If it is estimated that the car will be worth £6,000 in three years time when it is to be sold, then it could be charged to profit at (£15,000 − £6,000)/3 = £3,000 for each year of use.

Depreciation is discussed here as a prelude to our discussion about investment appraisal. The way companies accumulate funds with which to replace fixed assets is to charge depreciation as an overhead cost. In the case of our car this recovers £9,000 which together with the sale price of the used car generates a fund of £14,000 towards the purchase of a replacement. It is also possible that the sums so deducted from profit can be invested to offset inflation until the time comes to replace the asset.

Because depreciation is an estimate and is deducted from profit it has the effect of keeping the money available in the business. Many companies use the aggregate depreciation charged as the basis for the fund against which investment appraisal is done.

There are two main methods used to calculate depreciation. We will outline each in turn:

Straight line method

This charges an equal amount as depreciation for each year of the asset's expected life. It is called straight line method because if the annual amounts were plotted on a graph they would form a straight line. The formula is:

$$d = \frac{p - v}{n}$$

where d = annual depreciation, p = purchase price, v = residual value, and n = years of asset life. The method used for the car in the above example is straight line. It is a very popular and easy to use method.

Reducing balance method

In this method a fixed percentage is applied to the written down balance of the fixed asset. The formula for establishing this fixed percentage is:

$$r = 1 - \sqrt[n]{\frac{v}{p}}$$

where r = the percentage rate, n = number of years, v = residual value, and p = asset purchase price.

v must be a significant amount, otherwise the rate will be very large and somewhat meaningless.

Using our car example above the reducing balance rate, r, is:

$$r = 1 - \sqrt[3]{\frac{6,000}{15,000}} = 1 - 0.7368 = 26.31\%$$

Applying this, the depreciation pattern is as follows:

Purchase price		£15,000
1st year	26.31%	£3,946
Reduced balance		£11,054
2nd year	26.31%	£2,908
Reduced balance		£8,116
3rd year	26.31%	£2,135
Residual value (approx)		£5,981

The reducing balance system means that a greater sum is taken in earlier years, but reduces year on year since the percentage rate is applied to the reduced balance. This method is more complex but it is more logical because a new asset gives better service than an old asset and should suffer greater depreciation in earlier years. Certainly, with regard to motor cars the early years' depreciation is very heavy in relation to resale prices.

Investment appraisal

The process of investment appraisal is necessary in order to select the best projects for investment, whether they be replacement machines for the factory, office equipment or new cars for managers and sales representatives.

Major factors to consider are:

* risk and return
* time scale
* time value of money
* evaluating alternative choices.

The first one, that of risk and return is important. A good general rule is to expect a high return if the investment is risky. That is why interest rates on mortgages are less than interest rates on loans for consumer durables. A bank will regard loans against property as low risk because a house is a better security than a refrigerator if something goes wrong.

For commercial businesses some capital investment projects carry low risk, such as those which merely involve a replacement of a machine which is being used to make a product for which there is a regular demand. However, the risk is much greater for investment in a new factory for a range of new products. These different risk profiles have to be taken into account when comparing different projects for investment.

Time scale is important too. If the project is going to take 5 years to be completed, then many things could change before completion. Inflation, government action, market conditions, competitive pressures, etc. The investment decision must take account of all these risk factors as well and this is not an easy process, because of the need to forecast up to five years into the future.

Money has *time value*. This fact lies at the heart of the investment appraisal process. If you could choose to have £5 today or £5 in a year's time which would you prefer and why?

Hopefully you would take account of the risk of not getting your £5 in a year's time and would prefer to avoid risk and get it now. Also you should have thought about the fact that if you had the £5 now you could invest it for a year, so that it would be worth more than £5 in a years time.

The time value of money exists because of these two aspects, risk and return. Most people who decide to save money instead of spending it look for a secure and lucrative place to keep it until they wish to spend it.

If your risk preference is low you would probably put your money into a building society account. If you were more adventurous you might buy shares. The risks attached to buying shares are greater than a building society account, but the return may be greater.

The essence of the capital investment decision is no different to this. A company must decide whether or not to keep its money in the bank or invest in the business. If it chooses to invest in the business, the company must expect a better return than merely leaving it in the bank. It is for this reason that certain techniques for evaluating projects have arisen. There are numerous approaches, but the most theoretically sound method is called *Net Present Value* (NPV) technique.

Make or buy decisions

Make or buy decisions require the involvement of more than one functional area. They almost certainly require the involvement of the production function, production control and engineers from design and manufacturing.

Before an actual decision is made certain questions must be addressed:

(a) Could the item be made with existing facilities?
(b) If the answer to (a) is yes, is capacity adequate?

If the answer to (a) is no, then it indicates that new plant and equipment may be needed, or indeed that the company has no experience in this area of manufacture, which might therefore mean a move toward vertical integration, which involves a change to its basic business. The latter involves top level policy decisions and will not yet even be considered at the functional level at all. Theoretically any item, whether already manufactured or purchased outside can be reviewed for alternative sourcing, but in practice only those items will be considered for change that do not involve major changes to the business. Therefore, we will confine our analysis to decisions that can be made within the functional areas we described earlier.

Thus if a purchased item is being considered for manufacture then there must be suitable facilities available and sufficient capacity. If a manufactured item is being considered for purchase there must be alternative uses for the capacity or there should be an intention to reduce that capacity rather than leave it unused.

The other major factors to account for are:

- incremental costs of alternatives
- quality control problems
- multi-sourcing
- costs of tooling
- strategic importance.

What is *incremental cost*? This is the additional or unavoidable cost incurred for an item purchased outside or being manufactured inside. The point to note is that many overhead costs will be unaffected by the make or buy decision. Thus overhead or indirect costs are almost certainly irrelevant to the decision and only the direct or marginal costs are important. Thus comparison between the outside purchase price and the marginal or incremental manufactured price must be made to give a true picture.

If an item is being outsourced then quality control will need to be considered. Potential suppliers may not have in place suitable quality procedures or may not be part of any supplier quality assurance scheme. If multi-sourcing is necessary then this problem is even greater.

If an item for outside purchase is a special design to the company's own specification then capital investment costs will have to be accounted for. In these circumstances it is common for the customer and supplier to share the investment in tooling and this could be quite significant.

An item or component that is of strategic importance in the company's own manufacturing process presents risks if it is manufactured outside. This may be true if it is a crucial safety item or is something upon which other production processes depend. Thus a cessation of supply or quality problems could have serious consequences. Before arranging outside purchase these risks have to be evaluated and some form of risk reduction undertaken before committing to outside sourcing.

Inventory control

To avoid confusion we will use the term inventory control rather than stock control. In essence these terms are interchangeable.

Inventory control is of major importance in a manufacturing company. It has an impact upon production operations, marketing and finance. In many respects there are conflicting objectives between production, finance and marketing, thus a good inventory control system should achieve a satisfactory trade-off between these disparate needs.

For example, finance would seek to keep inventory levels as low as possible so that capital costs are reduced. Production might prefer large stocks of raw materials to ensure security of supply and to enable long production runs and smooth production. Marketing would like high stocks of finished goods in order to provide a high level of customer service, whereas finance would like to reduce the

---

Test your knowledge 1.16

1 Why are incremental costs the only relevant costs when considering make or buy decisions?
2 What other major factors affect the make or buy decision?
3 What major business functions are likely to be involved with make or buy decisions?

costs of warehousing, risk of pilferage and damage.

From this you can see that 'inventory' can be a stock of raw materials, components, work-in-process (semi-finished goods) or finished goods. You can also see that these different inventories represent the materials flow process as raw materials become components, which are put into assemblies and finally transformed into finished goods.

A periodic review system can work well if there is say a weekly order and delivery of goods. In which case, of course, individual item optimal review periods will be replaced by the convenience of standard review patterns for all goods, especially if they come from the same or few suppliers.

It can also work very well for inexpensive items like standard hardware (nuts, bolts, washers, clips, etc). In that case the amount ordered is based on physical inspection of bins at specified intervals. The bin size determines the target or maximum inventory, obviating detailed records.

Material requirements planning

We have already defined the basic principle of dependent demand systems, namely that demand is based on the demand for the final product, or on higher level assemblies. Thus in a manufacturing company raw materials, components and sub-assemblies are all dependent demand items, whilst the finished assembly or final stock is an independent demand item.

Inventories that conform to this definition should be controlled by a *Materials Requirements Planning* system or MRP for short. The system by its nature has to be fully computerized and as a consequence there are many proprietary systems on the market. Companies like Hewlett Packard, Honeywell and IBM market such packages. Other companies like Ford have designed their own systems.

The starting point for an MRP system is the *master schedule* that specifies the end items or final products and the quantities demanded. This could be for products as complex as cars or as simple as tables. The master schedule itself will be based on a sales forecast and may use sophisticated mathematical forecasting techniques.

From the master schedule is derived the demand for all subsidiary materials, components and assemblies. In effect the master schedule is 'exploded' into purchase orders or schedules for all parts and components required. This means that there has to be a bill of materials for each end product which specifies the quantities and types of parts needed.

The system has to take account of existing inventories of dependent demand items so that parts are ordered in the correct quantities. It must also consider order lead times for purchased parts and manufactured items, so orders may have to be issued in advance of actual usage to allow for the procurement time.

Test your knowledge 1.17

1 What items form part of the inventory held by an engineering company?
2 Explain why it is necessary to have some form of inventory control in an engineering company.
3 What is MRP?
4 What is a *bill of materials*?

Engineering and the environment

Environmental effects

Modern engineering processes and systems are increasingly designed and implemented to minimize environmental affects. Engineering companies must ensure that the negative effects of engineering activities on the natural and built environment are minimized. You need to be able to identify how individual engineering companies seek to do this, such as through:

- design of plant and products which optimizes energy use and minimize pollution
- good practice such as the efficient use of resources and recycling and the use of techniques to improve air and water quality
- management review and corrective action
- relevant legislation and regulations.

Material processing

Many engineering activities involve the processing of materials. Such materials may appear in the product itself or may be used in the manufacturing process. Some of these materials occur naturally and, after extraction from the ground, may require only minimal treatment before being used for some engineering purpose. Examples are timber, copper, iron, silicon, water and air. Other engineering materials need to be manufactured. Examples are steel, brass, plastic, glass, gallium arsenide and ceramic materials. The use of these materials produces effects; some beneficial, some not.

Economic and social effects stem from the regional wealth that is generated by the extraction of the raw material and its subsequent processing or manufacture into useful engineering materials. For example, the extraction of iron-ore in Cleveland and its processing into pure iron and steel has brought great benefit to the Middlesborough region. The work has attracted people to live in the area and the money they earn tends to be spent locally. This benefits trade at the local shops and entertainment centres and local builders must provide more homes and schools and so on. The increased numbers of people produces a growth in local services which includes a wider choice of different amenities, better roads and communications and arguably, in general, a better quality of life.

On the debit side, the extraction of raw materials can leave the landscape untidy. Heaps of slag around coal mines and steelworks together with holes left by disused quarries are not a pretty sight. In recent years much thought and effort has been expended on improving these eyesores.

Slag heaps have been remodelled to become part of golf courses and disused quarries filled with water to become centres for water sports or fishing. Disused mines and quarries can also be used for taking engineering waste in what is known as a landfill operation prior to the appropriate landscaping being undertaken.

Other potential problems can arise from having to transport the raw materials used in engineering processes from place to place. This can have an adverse affect on the environment resulting from noise and pollution.

Activity 1.13

Find out what happens to the domestic waste produced in your locality.

What items of domestic waste are recycled?

Is any of the waste burnt to produce useful heat?

If the waste is transported to another site for processing, where does it go and what processes are used?

What arrangements are there for disposing of hazardous waste?

Present your findings in the form of a brief word processed report. Illustrate your report with the aid of a simple flowchart.

The effects of waste products

Engineering activities are a major source of *pollutants* causing many types of *pollution*. Air, soil, rivers, lakes and seas are all, some where or other, polluted by waste gases, liquids and solids discarded by the engineering industry. Because engineering enterprises tend to be concentrated in and around towns and other built up areas, these tend to be common sources of pollutants.

Electricity is a common source of energy and its generation very often involves the burning of the *fossil fuels*: coal, oil and natural gas. In so doing, each year, billions of tonnes of carbon dioxide, sulphur dioxide, smoke and toxic metals are released into the air to be distributed by the wind. The release of hot gases and hot liquids also produces another pollutant; heat. Some electricity generating stations use nuclear fuel which produces a highly radioactive solid waste rather than the above gases.

The generation of electricity is by no means the only source of toxic or biologically damaging pollutants. The exhaust gases from motor vehicles, oil refineries, chemical works and industrial furnaces are other problem areas. Also, not all pollutants are graded as *toxic*. For example, plastic and metal scrap dumped on waste tips, slag heaps around mining operations, old quarries, pits and derelict land are all *non-toxic*. Finally, pollutants can be further defined as *degradable* or *non-degradable*. These terms simply indicate whether the pollutant decomposes or disperses itself with time. For example, smoke is degradable but dumped plastic waste is not.

Carbon dioxide

Carbon dioxide in the air absorbs some of the long-wave radiation emitted by the earth's surface and in so doing is heated. The more carbon dioxide there is in the air, the greater the heating or greenhouse effect. This is suspected as being a major cause of global warming causing average seasonal temperatures to increase. In addition to causing undesirable heating effects, the increased quantity of carbon dioxide in the air, especially around large cities, may lead to people developing respiratory problems.

Oxides of nitrogen

Oxides of nitrogen are produced in most exhaust gases and nitric oxide is prevalent near industrial furnaces. Fortunately, most oxides of nitrogen are soon washed out of the air by rain. But if there is no rain, the air becomes increasingly polluted and unpleasant.

Sulphur dioxide

Sulphur dioxide is produced by the burning of fuels that contain sulphur. Coal is perhaps the major culprit in this respect. High concentrations of this gas cause the air tubes in peoples' lungs to constrict and breathing becomes increasingly difficult. Sulphur dioxide also combines with rain droplets eventually to form sulphuric acid or *acid rain*. This is carried by the winds and can fall many hundreds of miles from the sulphur dioxide source. Acid rain deposits increase the normal weathering effect on buildings and soil, corrode metals and textiles and damage trees and other vegetation.

Smoke

Smoke is caused by the incomplete burning of the fossil fuels. It is a health hazard on its own but even more dangerous if combined with fog. This poisonous combination, called *smog*, was prevalent in the early 1950s. It formed in its highest concentrations around the large cities where many domestic coal fires were then in use. Many deaths were recorded, especially among the elderly and those with respiratory diseases. This lead to the first Clean Air Act which prohibited the burning of fuels that caused smoke in areas of high population. So-called smokeless zones were established.

Dust and grit

Dust and grit (or *ash*) are very fine particles of solid material that are formed by combustion and other industrial processes. These are released into the atmosphere where they are dispersed by the wind before falling to the ground. The lighter particles may be held in the air for a many hours. They form a mist, which produces a weak, hazy sunshine and less light.

Toxic metals

Toxic metals, such as lead and mercury are released into the air by some engineering processes and especially by motor vehicle exhaust gases. Once again the lead and mercury can be carried over hundreds of miles before falling in rain water to contaminate the soil and the vegetation it grows. Motor vehicles are now encouraged to use lead-free petrol in an attempt to reduce the level of lead pollution.

Ozone

Ozone is a gas that exists naturally in the upper layers of the earth's atmosphere. At that altitude it is one of the earth's great protectors but should it occur at ground level it is linked to pollution. *Stratospheric ozone* shields us from some of the potentially harmful excessive ultra–violet radiation from the sun. In the 1980s it was discovered that emissions of gases from engineering activities were causing a 'hole' in the ozone layer. There is concern that this will increase the risk of skin cancer, eye cataracts, and damage to crops and marine life.

At ground level, sunlight reacts with motor vehicle exhaust gases to produce ozone. Human lungs cannot easily extract oxygen (O_2) from ozone (O_3) so causing breathing difficulties and irritation to the respiratory channels. It can also damage plants.

This ground level or *tropospheric ozone* is a key constituent of what is called photochemical smog or summer smog. In the UK it has increased by about 60% in the last 40 years.

Heat

Heat is a waste product of many engineering activities. A typical example being the dumping of hot coolant water from electricity generating stations into rivers or the sea. This is not so prevalent today as increasingly stringent energy saving measures are applied. However, where it does happen, river and sea temperatures can be raised sufficiently in the region of the heat outlet to destroy natural aquatic life.

Chemical waste

Chemical waste dumped directly into rivers and the sea, or on to land near water, can cause serious pollution which can wipe out aquatic life in affected areas. There is also the long term danger that chemicals dumped on soil will soak through the soil into the ground water which we use for drinking purposes and which will therefore require additional purification.

Radioactive waste

Radioactive waste from nuclear power stations or other engineering activities which use radioactive materials poses particular problems. Not only is it extremely dangerous to people – a powerful cause of cancer – its effects do not degrade rapidly with time and remain dangerous for scores of years. Present methods of disposing of radioactive waste, often very contentious however, include their encasement in lead and burial underground or at sea.

Derelict land

Derelict land is an unfortunate effect of some engineering activities. The term derelict land may be taken to mean land so badly damaged that it cannot be used for other purposes without further treatment. This includes disused or abandoned land requiring restoration works to bring it into use or to improve its appearance. Land may be made derelict by mining and quarrying operations, the dumping of waste or by disused factories from by-gone engineering activities.

Environmental legislation

Engineering activities can have harmful effects on the physical environment and therefore on people. In order to minimize these effects, there is a range of legislation (rules and regulations) that all engineering companies must observe.

The appropriate United Kingdom Acts of Parliament include: Deposit of Poisonous Wastes Act, Pollution of Rivers Act, Clean Air

---

Test your knowledge 1.18

Name a pollutant that fits each of the following categories:

(a) toxic and degradable
(b) toxic and non-degradable
(c) non-toxic and degradable
(d) non-toxic and non-degradable.

Act, Environmental Protection Act, Health and Safety at Work, etc and the like. Additionally, not only are there local by-laws to be observed there are also European Union (EU) directives that are activated and implemented either through existing UK legislation in the form of Acts of Parliament or mandatory instructions called Statutory Instruments (SI).

New Acts and directives are introduced from time to time and Industry needs to be alert to and keep abreast of these changes. Typical of these new initiatives is the European Electromagnetic Compatibility (EMC) legislation. This states that with effect from 1st January 1996 it is a requirement that all products marketed must conform with the new legislation. This new EMC legislation, at last, officially recognizes the well known problem of unwanted electromagnetic wave radiation that emanates from most pieces of electrical equipment. The unwanted radiation can interfere with adjacent electronic equipment causing them damage or to malfunction.

In the case of UK Acts of Parliament, the above legislation is implemented by judgement in UK Courts of Justice in the normal manner but based on EU legislation, if more appropriate, or by judgement of the European Court of Justice.

The purpose of the of this legislation is to provide the following functions:

- *prevent* the environment being damaged in the first place;
- *regulate* the amount of damage by stating limits, for example, the maximum permitted amount of liquid pollutant that a factory may discharge into the sea;
- *compensate* people for damage caused, for example, from a chemical store catching fire and spreading wind borne poisonous fumes across the neighbourhood;
- *impose sanctions* on those countries or other lesser parties that choose to ignore the legislation;
- *define who is responsible* for compliance with legislation to persons who can be named and their precise area of responsibility documented.

Note: For the purposes of showing understanding of the above, you are *not* expected to have a detailed understanding of the various Acts however you *should be aware of the general provisions* of the legislation and what it is trying to achieve. Your school, college or local town library will be able to provide you with more details.

The effects of the above legislation on engineering activities has, in general, made them more difficult and more expensive to implement. A few simple examples of this follow.

- *Chemical factories* can no longer discharge their dangerous waste effluent straight into the river or sea without first passing it through some form of purification.
- *Coal fuelled power stations* must ensure that their chimney stacks do not pollute the neighbourhood with smoke containing illegal limits of grit, dust, toxic gases and other pollutants. A system of smoke filtration and purification must be (expensively) incorporated.
- *Motor car* exhaust gases must be sufficiently free of oxides of

nitrogen, carbon monoxide and other toxic gases. This can only be achieved by, among other things, replacing the crude petrol carburettor with a more sophisticated petrol injection system and fitting a catalyser in the exhaust pipe. All this has added to the price of the motor car and has made it more difficult for the DIY motorist to service his or her vehicle.

Activity 1.14

Write a report based on your investigations into the effects of ONE engineering activity from EACH category
in the range, *production*, *servicing* and *materials handling*, on the physical environment to include *human*, *natural* and *built*.

Examples of engineering production activities are:

- motor car manufacture,
- steel manufacture,
- coal mining.

Examples of engineering servicing activities are:

- motor car dealership garages,
- local council road maintenance depots,
- maintenance of electricity and gas supplies.

Examples of materials handling activities are:

- container handling terminals,
- moving cargo by rail and road,
- conveying goods on moving belts.

For the engineering activity you have chosen, your report should include a brief description of:

- the environmental effects of the materials used,
- the short-term and long-term environmental effects of any waste products,
- any environmental legislation effects giving specific examples.

Hints:

Make sure that your selected engineering activity gives you the opportunity to produce the necessary amount of evidence to demonstrate your competence and understanding.

You should approach this activity through case studies (e.g. those highlighted by court cases concerning failure to comply with legislation).

Finally, it is important to be clear about the difference *waste products* and *by-products*. The by-products from one process can be sold as the raw materials for other processes. For example, natural gas is a by-product of oil extraction and a useful fuel used in the generation of electricity. Waste products are those that cannot be sold and may attract costs in their disposal. Nuclear power station waste is a typical example.

Review questions

1   State FIVE key engineering industries. For each key industry sector, give an example of THREE engineered products or services.

2   Explain, briefly, how the aerospace industry contributes to the economy and employment in the UK.

3   Give THREE examples of service industries in which engineers are employed.

4   Define the following terms:

(a) *Gross National Product* (GNP)
(b) *Gross Domestic Product* (GDP).

5   Distinguish between *visible* and *invisible* exports.

6   Give TWO examples of invisible exports.

7   A company has the following departments: *Finance*, *Production*, *Purchasing* and *Personnel*. Which of these departments would be responsible for:

(a) employee welfare
(b) generating invoices for goods supplied
(c) obtaining supplies of raw materials
(d) preparing a manufacturing schedule.

8   Explain why *market testing* is often used before a new product goes into full-scale production. Which departments in an engineering company would be involved with market testing?

9   Describe the interface that exists between Finance and Production departments in a typical engineering company.

10   Define the following terms:

(a) *prime cost*
(b) *overhead cost*
(c) *variable cost*.

11   The following figures relate to the production of a small component:

Fixed cost: £20,000
Variable cost per unit: £4
Selling price: £6

Determine the breakeven point in sales turnover.

12   Explain the term *by-product* in relation to an engineering process.

13   Give an example of a by-product produced as a result of a typical engineering process.

14   Describe THREE management activities in a typical engineering company.

15   Draw an organization chart showing the structure of the *Finance* and *Accounting* functions of a typical engineering company.

16   Draw an organization chart showing the structure of the *Marketing* and *Sales* functions of a typical engineering company.

17   An engineering company wishes to develop a new product. Describe the main stages of the development process.

18   Explain, with the aid of relevant diagrams, the difference between *flat*, *heirarchical* and *matrix* structures used in engineering companies.

19   Identify the interfaces that exist between the following departments in a typical engineering company:

(a) *Marketing* and *Production*
(b) *Marketing* and *R&D*
(c) *R&D* and *Production*

20   Define the term *marginal cost*.

21   Sketch a graph showing how *costs* and *revenues* vary with *units of output*. Mark the *break even point(s)* on your graph.

22   Describe the major factors used in *investment appraisal*.

23   Explain what is meant by *incremental cost* in relation to a *make or buy* decision.

24   Explain what is meant by *material requirements planning* (MRP).

25   Describe, briefly, the effects of the following waste products produced as a result of engineering activities:

(a) carbon dioxide
(b) sulphur dioxide
(c) ozone
(d) smoke.

26   Engineering companies must organize their activities in such a way as to comply with relevant environmental legislation. Give TWO examples of typical engineering activities and the environmental legislation that relates to them.

27   Describe THREE ways in which a typical engineering company can seek to minimize the effects of its activities on the environment.

Unit 2 Application of new technology in engineering

Summary

New technology has been introduced into all sectors of engineering. Many engineering processes and engineered components rely heavily on the application of new technology. Indeed, many things that we take for granted would simply not be possible without the application of new technology. It should therefore not come as much of a surprise to find that that engineers need to be constantly aware of new technology and the likely impact that it will have on the future of engineering.

This chapter introduces a number of important aspects of new technology through a series of short case studies. Each case study focuses on a different product or service and shows how new technology has made the development possible. Each case study introduces one or more aspects of the application of the three main areas of new technology.

The eight case studies introduced in this chapter contain a number of activities for you to complete. These are designed to provide you with real-life examples of the development and application of new technology. They will also provide you with opportunities to develop your skills in gathering and using information. In order to complete the case study activities you will need to make use of your school or college library as well as other information sources such as CD-ROMs and the World Wide Web.

This unit has links with several other intermediate units, particularly those that involve the use of computer-aided design and making engineered products. You should look out for ways of using your knowledge of new technology in these units also!

Information and communications technology

Information and communications technology is not just about computers — it's about all aspects of accessing, processing and disseminating information. This revolution has come about as a result of the convergence of several different technologies, notably:

- computing (microprocessors, memories, magnetic and optical data storage devices, etc.)
- telecommunications (data communications, networking, optical fibres, etc.)

- software (optical character recognition, data transfer protocol, visually orientated programming languages, digital signal processing, etc.).

Case study	# Computer simulation

Computers provide us with powerful and cost-effective tools for designing, simulating, and analysing a wide variety of engineering problems. The advantage of computer simulation is that it can not only avoid expensive prototype costs but it can be performed away from the real world application and within a time frame that is most appropriate to the application concerned. The advantages of computer simulation are summarized below:

Changing the time frame

Simulation can take place within the most convenient time frame. In some cases simulation can take place in real-time (i.e. the computer performs its analysis within the same time frame as the real process that it is modelling). In many cases, the computer simulation is performed in a much shorter time scale (thus avoiding the need for lengthy delays during which very little may actually be happening in a real system). An example of using an accelerated time frame might be that of the simulation of a transport system where it is necessary to rapidly determine the impact of various controls and constraints on traffic flow. In other cases, the computer analysis may be performed over a much longer time scale than in the real world. This allows a more leisurely examination of events that take place very quickly. An example of using a longer time frame might be that of simulating the effect of a lightning strike on an electricity sub-station.

Capturing data for later analysis

Using a computer, data can be captured and recorded for further analysis at some later time. Computers can store large amounts of data and this can be easily retrieved at a later time for more extensive analysis. The results of several simulations using different parameters can then be compared and an optimum solution obtained. An example of a simulation with delayed data analysis might be that of a flight simulator in which the performance of a trainee pilot is later compared with that of an experienced pilot.

Simulating hazardous or dangerous processes

Processes that are potentially hazardous or dangerous can be simulated without any of the risks that would occur in a real world application. Some good examples can be found in the nuclear reprocessing industry where extremely hazardous materials are being handled on a daily basis.

Simulating costly processes

Processes that are very expensive can be analysed and perfected without wastage and without having to make a considerable cash outlay. Many examples can be found in the aerospace sector where aircraft and space vehicles can be extremely costly and where lives may be at risk where a failure occurs.

Typical examples of the use of computer simulation include:

Plant simulation

The use of computers to simulate the operation of manufacturing or processing plant is not particularly new. What is new is the use of relatively low-cost PCs to provide graphical displays of a process using icons and pictures of instruments and components representing those that would be present in a real plant.

Most of today's plant simulation software supports a graphical user interface and text displays have been almost entirely superseded. Using a graphical display it is easy to see whether an instrument or display is correct or whether there is a need to make an adjustment to one (or more) of the variables that are used to represent the real world values.

By making appropriate adjustments it is possible to simulate the failure of key components or sub-systems in order to ascertain whether a system can be safely shut down. It is also possible to experiment with demand and feed rates to check that the system is able to cope with the range of values that it might experience under both normal and abnormal circumstances.

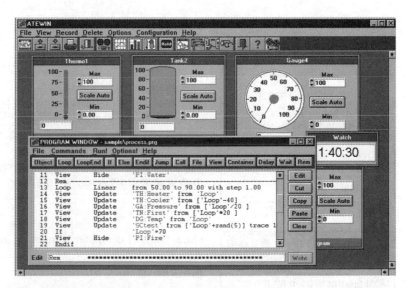

Figure 2.1 *Typical process simulation software which uses graphical displays of instruments and components*

Electronic circuit analysis

Computer packages that can be used for designing, simulating, and analysing analogue, digital and mixed (analogue/digital) electronic circuits are now commonplace. Circuits can be built either by using a *netlist* that describes all of the components and connections present in a circuit or by using an on-screen graphical representation of the circuit on test (which, in turn, generates a netlist or its equivalent).

Test your knowledge 2.1

State three advantages of using computer simulation in engineering.

Most programs that simulate electronic circuits use a set of *algorithms* that describe the behaviour of electronic components. The most commonly used algorithm was developed at the Berkeley Institute in the United States and it is known as SPICE (Simulation Program with Integrated Circuit Emphasis).

Results of circuit analysis can be displayed in various ways, including displays that simulate those of real test instruments (these are sometimes referred to as *virtual instruments*). A further benefit of using electronic circuit simulation software is that, when a circuit design has been finalised, it is usually possible to export a file from the design/simulation software to a PCB layout package. It may also be possible to export files for use in screen printing or CNC drilling. This greatly reduces the time that it takes to produce a finished electronic circuit.

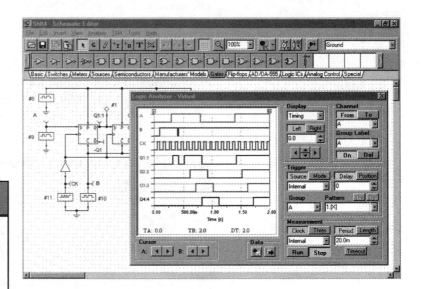

Figure 2.2 *Digital logic circuit simulation with virtual instrument logic analyser display.*

Test your knowledge 2.2

Explain the following terms in relation to electronic circuit design and simulation software:
(a) netlist
(b) SPICE
(c) virtual instrument.

Test your knowledge 2.3

Most programs used for the design and simulation of electronic circuits support a facility known as *file export*. Explain why this is a useful feature and suggest two applications for the exported files.

Other types of computer simulation include:

Magnetostatic simulation

Magnetic simulation is used to design or analyse variety of devices such as solenoids, electric motors, magnetic shields, permanent magnets, magnetic disk drives, and so forth. Generally the quantities of interest in magnetostatic analysis are magnetic flux density, field intensity, forces, torques, inductance, and flux linkage.

Electrostatic simulation

Electrostatic simulation is used to design or analyse variety of capacitive systems such as fuses, transmission lines and so forth. Generally the quantities of interest in electrostatic analysis are voltages, electric fields, capacitances, and electric forces.

Thermal simulation

Thermal analysis plays an important role in design of many different mechanical and electrical systems. Generally the quantities of interest in thermal analysis are temperature distribution, thermal gradients, and heat losses.

Current flow analysis

Current flow analysis is used to analyse a variety of conductive systems. Generally the quantities of interest in current flow analysis are voltages, current densities, electric power losses (Joule heat).

Stress analysis

Stress analysis plays an important role in design of many different mechanical and electrical components. Generally the quantities of interest in stress analysis are displacements, strains and different components of stresses.

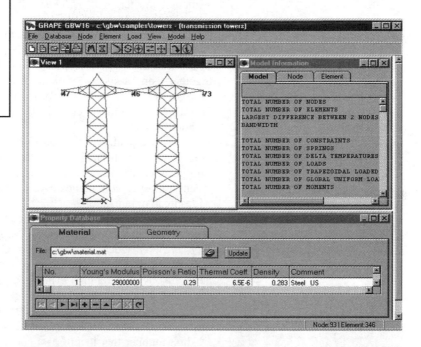

Figure 2.3 *Stress analysis software showing stresses and strains present in a complex structure*

> ## Test your knowledge 2.4
>
> Describe at least three typical engineering applications in which computer simulation is used.

> ## Test your knowledge 2.5
>
> Explain why computer simulation might be useful in each of the following engineering applications:
> (a) operating a remote controlled vehicle that will be used for deep sea exploration
> (b) designing a lightweight transportable bridge for use by armed forces in a battlefield situation.

Activity 2.1

Investigate the use of computer software that can be used to simulate electronic circuits. Packages that you might wish to investigate include Electronic Workbench, Tina Pro, and Micro-Cap. Find out what each package is capable of doing and produce a full specification for at least one of the packages. Present your findings in the form of a brief word processed report.

Activity 2.2

Investigate the use of computer software that can be used to analyse the stresses present in structures such as beams, bridges, and roofs. Obtain access to a full or evaluation copy of a stress analysis package and use it to analyse the stress present in a simply supported beam. Enter typical values corresponding to a simple real world application and obtain a print out of your results. Comment briefly on the use of the package and present your findings in the form of a brief word processed report with accompanying illustrations and screen shots.

Case study

The Internet and the World Wide Web

Although the terms *World Wide Web* and *Internet* are often used synonymously, they are actually two different things. The *Internet* is the global association of computers that carries data and makes the exchange of information possible. The *World Wide Web* is a subset of the Internet — a collection of inter-linked documents that work together using a specific Internet protocol called *hypertext transfer protocol* (HTTP). In other words, the Internet exists independently of the World Wide Web, but the World Wide Web can't exist without the Net.

The World Wide Web began in March 1989, when Tim Berners–Lee of the European Particle Physics Laboratory at CERN (the European centre for nuclear research) proposed the project as a means to better communicate research ideas among members of a widespread organization.

Web sites are made up of collections of Web pages. Web pages are written in *hypertext markup language* (HTML), which tells a *Web browser* (such as Netscape's Communicator or Microsoft's Internet Explorer) how to display the various elements of a Web page. Just by clicking on a *hyperlink*, you can be transported to a site on the other side of the world.

A set of unique addresses is used to distinguish the individual sites on the World Wide Web. An Internet Protocol (IP) address is a 4- to 12-digit number that identifies a specific computer connected to the Internet. The digits are organized in four groups of numbers (which can range from 0 to 255) separated by full stops. Depending on how an Internet Service Provider (ISP) assigns IP addresses, you may have one address all the time or a different address each time you connect.

Every Web page on the Internet, and even the objects that you see displayed on Web pages, has its own unique address, known as a *uniform resource locator* (URL). The URL tells a browser exactly where to go to find the page or object that it has to display.

Test your knowledge 2.6

What does each of the following abbreviations stand for?
(a) IP
(b) HTML
(c) URL
(d) ISP.

On-line services

On-line services can be defined as services that add value to the World Wide Web. Originally, these services built and maintained

Test your knowledge 2.7

Briefly explain each of the following terms:
(a) hyperlink
(b) hypertext
(c) hypertext transfer protocol
(d) hypertext markup language
(e) Web browser
(f) uniform resource locator.

Test your knowledge 2.8

Explain why sites on the Word Wide Web must all have unique addresses.

Test your knowledge 2.9

Name two popular Web browsers and explain what they are used for.

trunk networks that could be used by their customers. They also added their own *content* (such as news and weather reports, software libraries, etc.) to the Web. Users who are prepared to pay for the service can access this material.

Customers still pay for some on the on-line services but the trend, in recent years, has been to make added value services free. Currently the most popular on-line services are America Online (AOL), CompuServe, Prodigy, and the Microsoft Network (MSN). All of these services provide access to e-mail, support libraries, and on-line communities where people with similar interests can communicate for business or pleasure.

Search sites

Being able to locate the information that you need from a vast number of sites scattered across the globe can be a daunting prospect. However, since this is a fairly common requirement, a special type of site, known as a *search site,* is available to help you with this task. There are three different types of search site on the Web; *search engines*, *Web directories*, and parallel and *metasearch sites.*

Search engines such as Excite and HotBot use automated software called *Web crawlers* or *spiders.* These programs move from Web site to Web site, logging each site title, URL, and at least some of its text content. The object is to hit millions of Web sites and to stay as current with them as possible. The result is a long list of Web sites placed in a database that users search by typing in a keyword or phrase.

Web directories such as Yahoo and Magellan offer an editorially selected, topically organized list of Web sites. To accomplish that goal, these sites employ editors to find new Web sites and work with programmers to categorize them and build their links into the site's index.

To make things even easier, all the major search engine sites now have built-in topical search indexes, and most Web directories have added a keyword search.

Intranets

Intranets work like the Web (with browsers, Web servers, and Web sites) but companies and other organizations use them internally. Companies use them because they let employees share corporate data, but they're cheaper and easier to manage than most private networks — employees don't need any software more complicated or more expensive than a Web browser, for instance. They also have the added benefit of giving employees access to the Web. Intranets are closed off from the rest of the Net by *firewall* software, which lets employees surf the Web but keeps all the data on internal Web servers hidden from those outside the company.

Extranets

One of the most recent developments has been that of the *extranet.* Extranets are several intranets linked together so that businesses can share information with their customers and suppliers. Consider, for example, the production of a European aircraft by four major aerospace companies located in different European countries. They

Test your knowledge 2.10

Explain the meaning of each of the following terms:
(a) Internet
(b) intranet
(c) extranet
(d) firewall.

Test your knowledge 2.11

Explain the difference between a search engine and a Web directory.

Test your knowledge 2.12

Name three popular on–line service providers.

might connect their individual company intranets (or parts of their intranets) to each other, using private leased lines or even using the public Internet. The companies may also decide to set up a set of private newsgroups so that employees from different companies can exchange ideas and share information.

E-mail

Like ordinary mail, e–mail consists of a message, an address and a carrier that has the task of actually conveying the message from one place to another. The big difference is that e–mail messages (together with any attached files) are broken down into small chunks of data (called *packets*) that travel independently to their destination along with innumerable other packets travelling to different destinations.

The packets that correspond to a particular e-mail message may travel by several different routes and may arrive out of order and at different times. Once all the packets have arrived, they are recombined into their original form. This may all sound rather complicated but it is nevertheless efficient because it prevents large messages hogging all of the available bandwidth. To put this into context, a simple page of A4 text can be transferred half-way round the world in less than a minute!

Activity 2.3

Visit the Web site of Lansing Linde UK, a manufacturer of forklift trucks. Investigate the company and view some of the gallery of photographs that show the development of the company's forklift trucks. Search the site for information on its latest range of electric forklift trucks capable of handling loads from 1,000 kg to 8,000 kg. Write a brief report describing these trucks and make specific reference to features that:

- save energy
- ensure smoother driving
- protect the driver
- improve visibility
- facilitate turning and manoeuvring
- ensure stability.

The URL of the Lansing Linde Web site is:
http://www.lll.co.uk

Activity 2.4

Set up a Web-based e-mail account in your own name. Note down all of the steps that you took to open the account including details of any electronic forms that you had to complete. Present your findings in the form of a brief article for your local paper on how to open and use an e-mail account.

Activity 2.5

Use a search engine (such as Lycos or AltaVista) to locate information on MP3 players. Visit the first four sites displayed as a result of your search and note down the URL of each of these sites. Summarize the contents of each of the sites by writing a paragraph describing each site. Then rate each site on a scale of 1 to 10 on the basis of content, presentation, and ease of use.

Summarize your results in a table.

Now repeat the activity using a Web directory (such as Yahoo or Excite). Use the directory to navigate to four different sites giving details of MP3 players. Once again, note down the URL of each site, summarize its contents and rate it on a scale of 1 to 10 (again presenting your results in the form of a table). Compare these two search methods.

Case study Computer-aided engineering (CAE)

You will probably already know a little about computer-aided design (CAD) however this is just one aspect of computer aided engineering (CAE). Computer-aided engineering is about automating *all* of the stages that go into providing an engineered product or service. When applied effectively, CAE ties all of the functions within an engineering company together. Within a true CAE environment, information (i.e. data) is passed from one computer aided process to another. This may involve computer simulation, computer aided drawing (drafting), and computer aided manufacture (CAM).

The term, CAD/CAM, is used to describe the integration of computer–aided design, drafting and manufacture. Another term, CIM (computer integration manufacturing), is often applied to an environment in which computers are used as a common link that binds together the various different stages of manufacturing a product, from initial design and drawing to final product testing.

Whilst all of these abbreviations can be confusing (particularly as some of them are often used interchangeably) it is worth remembering that 'computer' appears in all of them. What we are really talking about is the application of computers within engineering. Nowadays, the boundaries between the strict disciplines of CAD and CAM are becoming increasingly blurred and fully integrated CAE systems are becoming commonplace in engineering companies.

We have already said that CAD is often used to produce engineering drawings. Several different types of drawing are used in engineering. Some examples of different types of CAD drawing are shown on the next page.

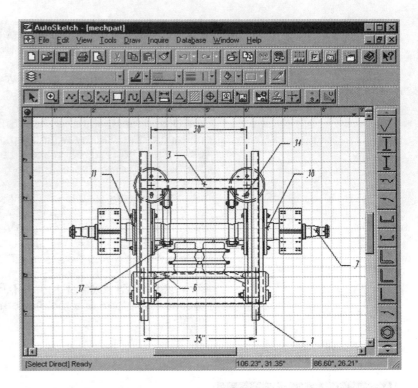

Figure 2.4 *A drawing of a mechanical part using a popular 2-D CAD package*

Figure 2.5 *A rendered 3-D representation of a car using a popular 3-D CAD package*

Computer aided manufacture (CAM)

Computer aided manufacture (CAM) encompasses a number of more specialized applications of computers in engineering including computer integrated manufacturing (CIM), manufacturing system modeling and simulation, systems integration, artificial intelligence applications in manufacturing control, CAD/CAM, robotics, and metrology.

Computer aided engineering analysis can be conducted to investigate and predict mechanical, thermal and fatigue stress, fluid flow and heat transfer, and vibration/noise characteristics of design concepts to optimize final product performance. In addition, metal and plastic flow, solid modeling, and variation simulation analysis are performed to examine the feasibility of manufacturing a particular part.

In addition, all of the machine tools within a particular manufacturing company may be directly linked to the CAE network through the use of centrally located floor managers which monitor machining operations and provides sufficient memory for complete machining runs.

Manufacturing industries rely heavily on computer controlled manufacturing systems. Some of the most advanced automated systems are employed by those industries that process petrol, gas, iron and steel. The manufacture of cars and trucks frequently involves computer-controlled robot devices. Industrial robots are used in a huge range of applications that involve assembly or manipulation of components.

The introduction of CAD/CAM has significantly increased productivity and reduced the time required to develop new products. When using a CAD/CAM system, an engineer develops the design of a component directly on the display screen of a computer. Information about the component and how it is to be manufactured is then passed from computer to computer within the CAD/CAM system. After the design has been tested and approved, the CAD/CAM system prepares sequences of instructions for computer numerically controlled (CNC) machine tools and places orders for the required materials and any additional parts (such as nuts, bolts or adhesives). The CAD/CAM system allows an engineer (or, more likely, a team of engineers) to perform all the activities of engineering design by interacting with a computer system (invariably networked) before actually manufacturing the component in question using one or more CNC machines linked to the CAE system.

Test your knowledge 2.13

What do each of the following abbreviations stand for?
(a) CAD
(b) CAM
(c) CAD/CAM
(d) CIM
(e) CNC
(f) CAE.

Test your knowledge 2.14

Explain the advantage of using CAD/CAM in the design and manufacture of an engineered product.

Activity 2.6

Use a CAD package, such as Autosketch, to produce a 2-D plan of the workshop area in your school or college. Choose an appropriate scale and ensure that your plan is fully dimensioned. Mark on your plan the location of machine tools and other equipment available. Print or plot the finished plan and ensure that you have adopted the recommended style, framing and presentation conventions used in your school or college.

Telecommuni-cations

Telecommunications provides the infrastructure that allows information to be transferred from place to place. The term includes communications by conventional wires (or *lines*), cables, radio, microwave and optical technology.

Case study   Digital audio broadcasting

Digital audio broadcasting (DAB) is arguably the biggest revolution in radio since the early days of broadcasting. DAB shows enormous potential and manufacturers are working with key industry groups in order to bring this new technology to the user. Much of the existing analogue broadcast radio service is likely to be phased out within 10 to 15 years and this has added a degree of urgency to getting the new digital broadcasting services up and running.

The first technology to be used for radio broadcasting was based on the amplitude modulation (AM) of radio signals in the long and medium wave bands. This technology was employed in the first 50, or so, years of radio broadcasting and it was sufficient (in terms of quality) to develop an expanding and enthusiastic radio audience in virtually every country of the world. National coverage can be easily provided by one or more high-power AM transmitters operating in the long and medium wave bands. Furthermore, provided that high–quality audio is not required, the spectrum requirements are relatively undemanding.

FM (frequency modulation) broadcasting started some 50 years ago in an effort to overcome the vagaries of propagation at the lower radio frequencies and also to very significantly reduce problems associated with amplitude noise and propagation disturbances. The FM broadcasting service was essentially a local service, requiring multiple transmitters operating on different frequencies in order to provide national coverage. It was (and still is) very inefficient in terms of its use of radio frequency spectrum.

Conventional AM and FM radio uses signals that are, between certain limits, continuously variable. These analogue signals are prone to various forms of noise and require that special measures (such as compression) are used in the broadcasting process in order to improve signal-to-noise ratios. Ignition and other electrical interference, fading and distortion caused by multi-path reception are further causes of distortion which, whilst much reduced using FM broadcasting techniques, can seriously degrade the quality of the received signal.

Digital radio offers many advantages, including the ability to remain reasonably impervious to the effects of amplitude noise and other signal disturbances as well as the ability to re–use spectrum space (using techniques that are not possible with conventional analogue broadcasting).

The technical specifications for digital audio broadcasting were originally developed by a project team known as 'Eureka 147' and, whereas this was originally a European group (with representatives from a number of interested bodies including broadcasting authorities), it now has members from all over the world. The World DAB organization is dedicated to encouraging international co-operation and co-ordination between sound and data broadcasters, network providers, manufacturers, governments and other official bodies in order to gain consensus for the smoother introduction of digital audio broadcasting worldwide.

Conventional terrestrial broadcasting (using transmitter sites all over the UK) is not the only way of receiving DAB. Digital radio can now be received in more than two million homes in the UK via digital satellite broadcasting in the L-band and a set-top box.

Future digital radio receivers will be fitted with pixel matrix displays that will permit the display of graphical symbols and simple pictures. The ability to display scrolling text messages, simple graphics or road maps could be invaluable in many situations.

In DAB, the audio (and data) signals broadcast in a digital radio system must be converted to digital format before they can be broadcast. At the receiving end, the digital information recovered from the received radio wave must be converted back into analogue form. Digital modulation techniques must be employed at the transmitter whilst digital demodulation must be employed at the receiver.

A number of digital radio programmes and services are conveyed in one contiguous block of radio frequencies. This band of frequencies is known as the *multiplex*. The multiplex allowing broadcasters to group together a number of programmes and additional data services that are all transmitted within a frequency channel. Each multiplex can carry a mixture of stereo and mono broadcasts and data services.

Since there is a trade-off between bandwidth and audio quality (in terms of the highest audio frequency that can be transmitted) the fewer the number of services carried, the higher the audio quality that can be allocated to each service. Multiplexing allows digital broadcasters to exploit the trade-off between bandwidth and audio quality and they thus have a high degree of flexibility in how they use their allocated bandwidth.

A DAB multiplex is made up of 2,300,000 bits (*binary digits*) that are used for carrying audio, data and error correction. Of these, approximately 1,200,000 bits are used for the audio and data services. In order to ensure efficiency, a different number of bits may be allocated to each service.

The seven multiplexes in the UK have been allocated as follows:

- one multiplex for BBC national radio
- one multiplex for national commercial radio (awarded to Digital One)
- five multiplexes for local radio in England and national stations in Scotland, Wales and Northern Ireland (BBC and commercial).

The UK Government has allocated the seven multiplexes to Band III (the range of frequencies formerly occupied by 405-line ITV signals) in the frequency range from 217.5 MHz to 230.0 MHz. The BBC national multiplex is located at 225.648MHz. Each multiplex can carry a mixture of stereo and mono broadcasts and data services, the number of each is determined by the audio quality required but services can be varied throughout the day according to programme schedules.

Apart from the obvious improvement in sound quality resulting from an improvement in noise reduction, digital radio offers a number of other significant advantages over conventional analogue-based broadcasting technologies. These include:

- *Enhanced services* Digital radio makes far more efficient use of the available bandwidth than do conventional broadcasting technologies. This allows broadcasters to provide additional services as well as providing a wider range of programmes. Most UK listeners will, for example, be able to receive more than 12 national and 12 regional and local services broadcast both by the BBC and by commercial broadcasting companies. Additional data-related services will include the provision of news, traffic information, sports results as well as selective programme content which will offer listeners more choice and customizable options.

- *Reduced interference* Provided that received signal levels are above a certain minimum threshold, digital radio signals remain virtually impervious to noise and amplitude borne interference. Common problems associated with the propagation of radio waves (such as fading on HF and aircraft reflection on VHF) have negligible effect on digital signals.

- *Simplified tuning requirements*   Digital radio receivers (particularly those designed for portable and in-car use) require minimal re-tuning when the receiver is moved from place to place. The user can simply select a particular programme without having to alter the tuning to locate the most favourable transmitter source.

- *Data displays* Digital radio receivers will be able to accept data as well as conventional audio signals. This will allow them to display scrolling text and graphical information using a built-in LCD display. It will, for example be possible to display news headlines, sports results and half-time scores as well as programme schedules, details of artists, etc.

- *Improved quality and reduced distortion*   With digital radio there is no need for the audio compression that has to be applied to signals broadcast using conventional analogue broadcasting technology. The removal of audio compression is instrumental in producing audio signals which very closely resemble the original source material. A comparison that is sometimes made when attempting to describe the improvement in signal quality is the difference that exists between vinyl records and compact disks. In reality, and because of the vagaries of the transmission medium employed, the improvement in quality between digital and analogue radio is *much* greater!

Digital audio broadcasting (DAB) is not just one technology but the combination of several technologies, notably:

- *MUSICAM* MUSICAM is a digital *compression* system that is used to reduce the vast amount of digital information required to be broadcast to a manageable amount. It does this by eliminating from the broadcast channel signals that will not be perceived by the listener (e.g. very quiet sounds that are masked by other louder sounds).

- *COFDM* COFDM (or *coded orthogonal frequency division multiplexing*) ensures that signals are received reliably, even in environments normally prone to interference. Using a precise mathematical transformation, the MUSICAM signal is split across 1,536 different carrier frequencies, and also divided in time. This process ensures that even if some of the carrier frequencies are affected by interference, or the signal disturbed for a short period of time, the receiver is still able to recover the original sound. The interference that disturbs conventional FM radio signals (e.g. multi-path reception and aircraft reflection) is significantly reduced by applying COFDM technology. COFDM also allows spectrum re-use so that the same broadcast frequency can be used across the entire country.

- *SFN* In digital radio broadcasting, the same frequency block of spectrum is effectively re-used throughout the service area in what is known as a *single frequency network* (SFN). Within the SFN, all the transmitters use the same frequency to broadcast the same digital radio signal. Hence there is no need to re-tune a receiver that is moved from place to place. Once selected, the required programme remains tuned.

- *TPEG* TPEG is a set of traffic information protocols. The Transport Protocol Experts Group (TPEG) was commissioned by the European Broadcasting Union's EBU Broadcast Management Committee to develop a new protocol for traffic and travel information for use in the multimedia broadcasting environment. TPEG technology is expected to revolutionize traffic information services by providing a personal travel service that allows a receiver to only use the traffic news that applies to the area in which it is currently located. This digital service will provide traffic information on demand and in much greater detail than ever previously possible. In addition to providing up to the minute information such as delays and accidents, the data can be translated into a wide range of formats including text or graphical display and voice synthesized. In future it is also possible that data could be available in several different languages. TPEG also offers the possibility of selecting information that is relevant to an individual motorist. Not only can it help in the selection of routes to avoid delays caused by traffic congestion or road works but it can also suggest alternative routes. Integration of digital radio receivers using TPEG with vehicle navigation systems (such as those based on RDS and Navstar satellites) is a further possibility that we are likely to see in the near future.

Each service component within a DAB signal carries an audio or data part of a service. Information that links the various components of a particular service is carried in the *fast information channel* (FIC). Broadcasting organizations can link a number of different *service components* together to produce a complete services. One component must be defined as the *main component*, the others are called *secondary components*. For example, the BBC's principal radio services (Radios 1, 2, 3, 4 and 5 Live) will be transmitted as main service components. Each of the principal services may have secondary components. For example, Radio 5 Live Plus could be introduced into the multiplex to provide additional coverage of a major sports event.

Activity 2.8

Use the Internet or other information sources to obtain information and specifications on at least three currently available DAB receivers (these are manufactured and/or distributed in the UK by companies like Arcam, Clarion, Cymbol, Grundig, Kenwood, Pioneer, Sony, Tag McLaren, and Technics). Present your finding in the form of a series of word processed data.

Test your knowledge 2.15

What does each of the following abbreviations stand for?

(a) DAB
(b) AM
(c) FM
(d) SFN
(e) COFDM
(f) TPEG
(g) RDS
(h) FIC.

Activity 2.9

Write a brief word-processed article for your local paper suitable for non-technical readers explaining the advantages of DAB.

Activity 2.10

Use the Internet to obtain information on the DAB services that are currently available from the BBC. Explain how these services improve on services based on conventional analogue broadcasting technology. Present your findings in the form of a brief word–processed fact sheet.

Test your knowledge 2.16

Explain why VHF FM broadcasting provides better quality reception than conventional AM broadcasting. Give at least *two* reasons.

Test your knowledge 2.17

Some DAB receivers provide coverage of signals in L-band (with a typical tuning range of 1,452 MHz to 1,492 MHz) as well as Band III (with a typical tuning range of 174 MHz to 240 MHz). Explain why this is.

Activity 2.11

Future digital radio receivers are likely to be very different from conventional radio receivers. Explain why this is and suggest ways in which this new technology will provide additional services of specific benefit to the motorist. Present your findings in the form of a brief class presentation with appropriate visual aids.

New materials and components

Ongoing developments in materials and materials processing has made possible a number of exciting new products. The case study in this section is about a group of materials commonly referred to as plastics but more correctly called polymers.

Case study

Polymers

Polymers are large molecules made of small, repeating molecular building blocks called *monomers*. The term polymer is a composite of the Greek words *poly* and *meros,* meaning 'many parts'. The process by which monomers link together to form a molecule of a relatively high molecular mass is known as *polymerization.*

Polymers are found in many natural materials and in living organisms. For example, proteins are polymers of amino acids, cellulose is a polymer of sugar molecules, and nucleic acids such as deoxyribonucleic acid (DNA) are polymers of nucleotides. Many man-made materials, including nylon, paper, plastics, and rubbers, are also polymers. Plastics are often divided into two main groups; *thermoplastics* and *thermosetting plastics* (or *thermosets*).

There is an important difference between these two classes of material. Because themosetting plastics undergo a chemical change during moulding, they cannot be softened by reheating. Thermoplastics, on the other hand, become soft whenever they are reheated. Thermoplastics are softer and more pliable than thermosetting plastics which tend to be hard and brittle.

The chemical change experienced by thermosetting plastic materials (referred to as called *curing*) is brought about from the temperature and pressure that is applied during the moulding process. The properties of some common thermoplastics are shown in Table 2.1.

Table 2.1 *Properties of some common thermoplastics*

Material	Relative density N/mm^2	Tensile strength J	Elongation %	Impact strength	Max. service temperature °C
Polyamide (nylon)	1.14	50–85	60–300	1.5–15	120
Polythene	0.9	30–35	50–600	1–10	150
Polypropylene	1.07	28–53	1–35	0.25–2.5	65–85
Polystyrene	1.4	49	10–130	1.5–18	70
Polyvinyl chloride (PVC)	1.3	7–25	240–380	–	60–105
Teflon (PTFE)	2.17	17–25	200–600	3–5	260

Table 2.2 *Properties of some common thermosetting polymers*

Material	Relative density N/mm^2	Tensile strength J	Elongation %	Impact strength	Max. service temperature °C
Epoxide	1.15	35–80	5–10	0.5–1.5	200
Urea formaldehyde	1.50	5–75	1.0	0.3–0.5	75
Polyester (alkyd resin)	2.00	20–30	0	0.25	150
Silicone	1.88	35–45	30–40	0.4	450

Activity 2.12

A company called DuPont invented Mylar polyester film in the early 1950s. Investigate the development and use of Mylar in the manufacture of electronic components. Write a brief word processed technical report giving your findings.

New materials and components

The information revolution has largely been made possible by developments in electronics and in the manufacture of integrated circuits in particular. Ongoing improvements in manufacturing technology have given us increasingly powerful integrated circuit chips. Of these, the single-chip microcomputer (a chip that performs all of the essential functions of a computer) has been one of the most notable developments. Single-chip microcomputers are found in a vast range of products, from domestic central heating controllers to complex engine management systems. The single-chip microcomputer is the subject of our next case study.

Case study

The single-chip microcomputer

A single-chip microcomputer (MCU) is a single chip of silicon that performs all of the essential functions of a computer including *central processor unit* (CPU), *input/output ports* (I/O), *read-only memory* (ROM), and *random access memory* (RAM) on a single silicon chip.

As in any microprocessor based system, the CPU controls the operation of the system, fetching and executing instructions and performing arithmetic and logical operations. The CPU also contains some general purpose registers that are used for storing a limited amount of data during processing. Where a single-chip microcomputer is used in a control system it is often referred to as a *microcontroller* or *microcontroller unit* (MCU). The data sheet for a simple 4-bit single-chip microcomputer is shown in Figure 2.6. A typical system using a single-chip microcomputer is shown in Figure 2.7.

HMCS45C,HMCS45CL
HD44820,HD44828
4-BIT Single-Chip Microcomputer

The HMCS45C is the CMOS 4-bit single chip microcomputer which contains ROM, RAM, I/O and Timer/Event Counter on single chip. The HMCS45C is designed to perform efficient controller function as well as arithmetic function for both binary and BCD data. The CMOS technology of the HMCS45C provides the flexibility of microcomputers for battery powered and battery back-up applications.

- **FEATURES**
- 4-bit Architecture
- 2,048 Words of Program ROM (10 bits/Word)
  128 Words of Pattern ROM (10 bits/Word)
- 160 Digits of Data RAM (4 bits/Digit)
- 44 I/O Lines and 2 External Interrupt Lines
- Timer/Event Counter
- Instruction Cycle Time; HMCS45C: 10μs
  HMCS45CL: 20μs
- All Instructions except One Instruction; Single Word and Single Cycle
- BCD Arithmetic Instructions
- Pattern Generation Instruction
  − Table Look Up Capability −
- Powerful Interrupt Function
  3 Interrupt Sources
  ┌─ 2 External Interrupt Lines
  └─ Timer/Event Counter
  Multiple Interrupt Capability
- Bit Manipulation Instructions for Both RAM and I/O
- Option of I/O Configuration Selectable on Each Pin; Pull Up Resistor or CMOS or Open Drain
- Built-in Oscillator (Resistor or Ceramic Resonator)
- Built-in Power-on Reset Circuit (HMCS45C only)
- Low Operating Power Dissipation; 2mW typ.
- Stand-by Mode (Halt Mode); 50 μW max.
- CMOS Technology
- Single Power Supply, HMCS45C: 4.5~5.5V
  HMCS45CL: 2.5~5.5V
- 54-pin Flat Plastic Package (FP-54) or 64-pin Dual-in-line Plastic Package (DP-64)

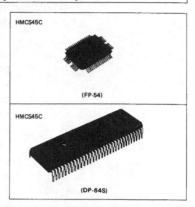

HMCS45C

(FP-54)

HMCS45C

(DP-64S)

- **PIN ARRANGEMENT**

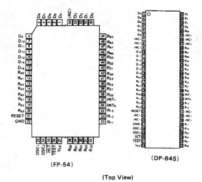

(FP-54)

(DP-64S)

(Top View)

Figure 2.6 *Data sheet for a simple 4-bit single-chip microcomputer*

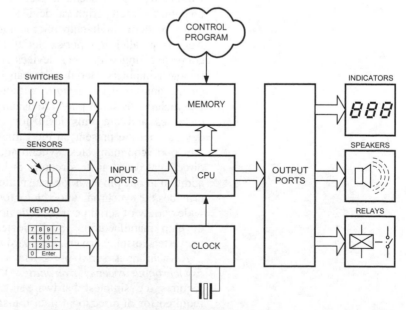

Figure 2.7 *A typical system based on a single-chip microcomputer*

A typical single-chip microcomputer is capable of executing 500,000 instructions per second. Whereas this speed is not comparable with the processing speed offered by the latest Pentium processors found in modern PCs, it is more than adequate for all but the most demanding of control system applications.

To be useful in a practical control system application, a single-chip microcomputer must have more than just a powerful CPU, a program, and a memory. In addition, it must contain circuitry that will allow the chip to be interfaced with the outside world. With a basic single-chip microcomputer this facility may be provided by incorporating a number of input/output (I/O) ports into the chip (see Figure 2.8). Each port has a number of lines (usually arranged in groups of four or eight) that can be configured as either inputs or outputs. Lines configured as inputs can accept digital signals from sensors and input transducers (including switches, buttons and keypads) whilst those configured as outputs can be connected to output devices such as audible transducers (speakers), light emitting diodes, relays, and solid-state switches. These, in turn can be connected to motors, fans, pumps and other external hardware.

In more complex single-chip microcomputers (and, more commonly, in microcontrollers) *on-chip peripherals* are incorporated within the chip in order to extend its capability. An MCU with on-chip peripherals can do a great deal more than one that has only general-purpose I/O (input/output) ports. Peripherals serve specialized needs and also reduce the processing load on the CPU. A prime example of an on-chip peripheral is an analogue-to-digital converter (ADC). Such a device allows analogue input signals (such as those from a temperature sensor) to be connected directly to the MCU.

Peripherals normally use *interrupt* signals to alert the CPU to the need to attend to some condition that requires its attention. This allows the CPU to get on with more demanding and important tasks without having to spend a great deal of its time controlling the operation of each peripheral device.

Many modern single-chip microcomputers have serial I/O ports as well as parallel I/O ports. Serial I/O ports provide a means of communicating with other devices such as external peripherals and remote computers. Serial ports can be either *synchronous* (using a clock signal) or *asynchronous* (without a clock signal).

Synchronous serial ports are usually only suited to relatively short distances and are thus ideal for connecting external peripheral devices that are present on the same printed circuit board. Because the port communicates synchronously with other devices, bi-directional data transfers require at least three pins on the MCU. In addition to one pin each for transmitted and received data, a third pin provides the synchronization clock for the communicating devices. A wide variety of serial peripheral devices are available from a number of chip manufacturers. Such devices include analogue-to-digital converters, display drivers, keyboard interfaces, etc.

Asynchronous serial ports use a device known as a *universal asynchronous receiver transmitter* (UART). This type of serial port requires the simplest hardware interface and only two pins are required for bi-directional data transfers. Data is transmitted out of the MCU on one pin and data is received by the MCU on the other pin. Each piece of data transmitted or received by the serial

Test your knowledge 2.20

What do each of the following abbreviations stand for?
(a) MCU
(b) CPU
(c) RAM
(d) ROM
(e) I/O
(f) UART.

Test your knowledge 2.21

State the functions of the CPU in a single-chip microcomputer.

Test your knowledge 2.22

Explain the function of an I/O port on a single-chip microcomputer.

communications port has a *start bit*, several *data bits*, and a *stop bit*. The start and stop bits are used to synchronize the two devices that are communicating. This type of serial interface is conventionally used to communicate over relatively long distances with a remote computer.

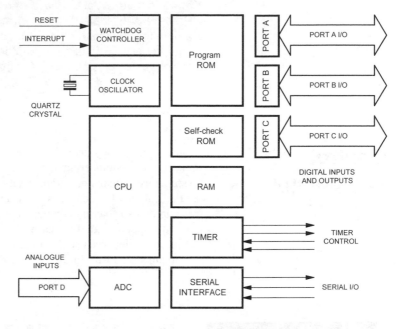

Figure 2.8 *Simplified internal architecture of an MCU*

Activity 2.13

Use the Internet or other information sources to obtain manufacturers' data on at least three comparable single-chip microcomputers from different manufacturers. Use a comparison chart to summarize the main features of each device and present your findings in the form of a short class presentation using appropriate visual aids.

Activity 2.14

The die used to produce a single–chip microcomputer is cut from a thin wafer of silicon on which a number of identical chips are fabricated. Draw a series of simple sketches that illustrate the main stages in the process of chip manufacture, starting with a cylinder of pure silicon and ending with a chip encapsulated in a package with pins that allow it to be connected to a socket mounted on a printed circuit board.

Activity 2.15

In recent years, PIC chips have begun to rival more single-chip microcomputers in simple control system applications. Use your library, the Internet and manufacturers' data sheets to investigate PIC devices. Present your findings in the form of a word processed article for publication in the electronic press. Your article should describe a generic PIC device and should include examples of at least two PIC applications.

Control technology

Control technology provides us with a means of automating a large number of engineering and manufacturing processes. Control systems use sensors and transducers to determine what is happening at the output of a control system and a processing device (often a microprocessor-based computer) to make decisions about what should happen next. Our case study introduces the programmable logic controller – a widely used solution to the problem of how to automate an industrial process.

Case study   Programmable logic controllers (PLC)

Programmable logic controllers (PLCs) are the unsung heroes of a huge variety of industrial processes – from brewing beer to handling the baggage in the world's busiest airports. Actually, there's nothing particularly new about the concept of a programmable controller (General Motors first proposed the concept nearly 30 years ago). Recently, however, technology has moved on apace and PLCs are becoming more sophisticated and indispensable than ever.

PLCs provide a rugged, low-cost yet versatile means of controlling a huge variety of industrial processes. When compared with other methods (such as the use of industrial PCs or dedicated microcontrollers) PLCs offer the following notable advantages:

- easily programmed and reprogrammed
- rugged enough to withstand a harsh electrical and mechanical environment
- modular – permitting easy removal and replacement in the event of failure
- flexible – allowing rapid reconfiguration of both hardware and software
- reliable and cost effective.

The first ever specification for a PLC was produced by General Motors in 1968 where a group of production engineers and designers were searching for a flexible solution to the need to control a number of complex linked manufacturing processes. Today, PLCs are manufactured by company's such as Allen–Bradley, Mitsubishi,

Siemens, General Electric, and Hitachi.

As with the development of personal computers, a number of significant events have determined the way that technology has changed over the decades. The following milestones are particularly noteworthy:

1968 First attempt at specifying a PLC

1969 First PLC based on discrete logic architecture and 1K of memory

1971 First microprocessors become available with a full range of arithmetic and logic operations, 256 input/output (I/O) addresses and an instruction set to facilitate programming

1977 PLCs use microprocessor central processing units (CPU)

1979 Larger memories become available (4K, 8K and 16K) and up to 1K of I/O

1981 Improved I/O based on programmable very large scale integrated (VLSI) devices

1983 Low-cost PLCs become available from several manufacturers

1985 PLCs first become networked

1987 Improved programming devices become available using portable and laptop PCs

1989 Several Japanese manufacturers become actively involved with PLCs

1991 Further improvements in networking permit large factory systems

The block schematic of a PLC is shown in Figure 2.9. The heart of the PLC is a central processing unit (invariable a microprocessor) and its immediate support devices, ROM, RAM and timers/counters. In addition, two dedicated areas of RAM (known as the process input image and process output image) help keep track of the state of all of the system's inputs and outputs (more about this later). In addition, most PLCs incorporate some on-board I/O which allows the controller to be used in simple applications without the need for any external I/O modules.

A series interface provides a means of programming the PLC (either by means of a hand-held programmer or by downloading programs from a remote PC). Alternatively, a specialized programming interface may be provided leaving the serial port free to permit data transfer to/from a remote computer.

External I/O is made possible by means of a number of modules connected to a common data path (or 'bus') linked to the controller. These modules contain the necessary interface circuitry (including level-shifting, optical-isolation, analogue-to-digital conversion and digital-to-analogue conversion, as appropriate).

A typical PLC arrangement based on a Siemens S5-95U PLC is shown in Figure 2.10. This shows a number of features in common with all PLCs – notably the modular construction based on expandable bus units.

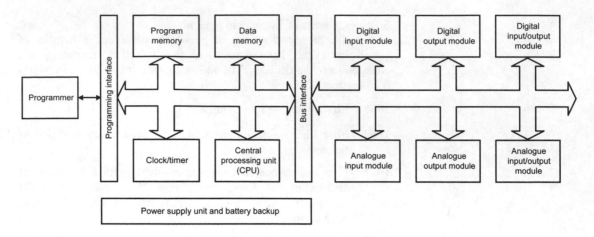

Figure 2.9 *Block schematic of a PLC*

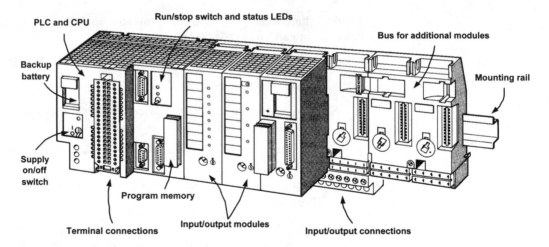

Figure 2.10 *The Siemens S50-95U PLC*

I/O modules

Although most PLCs have some form of on-board I/O, this is often somewhat limited (typically it is limited to a few digital I/O lines plus one or two analogue inputs). Most 'real-world' applications require a PLC to be concerned with a much larger number of inputs and outputs and thus additional I/O must be provided.

The following types of I/O module are available:

- digital input and digital output modules
- analogue input and analogue output modules
- timer modules
- counter modules
- comparator modules
- simulator modules
- diagnostic modules
- other modules including, stepper motor controllers, communications modules, etc.

Test your knowledge 2.23

Name three PLC manufacturers.

Test your knowledge 2.24

State three typical applications in which PLCs are used.

Test your knowledge 2.25

State three advantages of PLCs over other types of control system.

Activity 2.16

To provide a very high degree of isolation between the input and output circuits and the PLC, I/O modules often use *optocouplers*. Find out how an optocoupler works. Illustrate your answer with a sketch showing the internal circuitry of a typical optocoupler. Use an electronic component supplier's catalogue to determine the maximum voltage isolation for a typical optocoupler. Present your findings in the form of a word processed data sheet suitable for electronic hobbyists.

Activity 2.17

Analogue I/O modules provide a means of connecting analogue sensors and transducers to a PLC. Analogue output modules use digital to analogue converters (DAC) to convert digital signals from the microprocessor I/O to analogue signals that are required at the output of the PLC. Find out how a DAC works. Illustrate your answer with a block diagram. Present your findings in the form of a class presentation with appropriate visual aids.

Investigating new products

The engineering sectors and typical new technology products with which you need to be familiar are as follows:

Aerospace
New passenger and military aircraft, satellites, space vehicles, missiles, etc. from companies such as British Aerospace, Westland and Rolls-Royce.

Electrical and electronic
Electric generators and motors, consumer electronic equipment (radio, TV, audio and video), power cables, computers, etc. produced by companies such as GEC, BICC and ICL.

Mechanical
Bearings, agricultural machinery, gas turbines, machine tools and the like from companies such as RHP, GKN and Rolls-Royce.

Telecommunications
Telephone, radio and data communications equipment, etc. from companies like Nokia, GEC, Plessey and British Telecom.

Within the above sectors you need to be able to identify and investigate a variety of products that are based on the use of new technology. Our final case study provides you with an example of the application of several new technologies that you will already be familiar with!

Test your knowledge 2.26

To which of the above sectors do each of the following new technology products belong:
(a) MP3 players
(b) DAB receivers
(c) High-temperature Mylar dielectric capacitors
(d) CNC machine tools.

Activity 2.18

A group of Estate Agents has asked you to carry out some research for them based on engineering companies in your area. They are particularly interested in companies that are active in the field of new technology.

(a) Identify two engineering firms (taken from different sectors) that are involved in the design, manufacture or use of new technology products.
(b) List the new technology products that each company is involved with.
(c) Identify the engineering sector within which the company is working.
(d) Briefly describe the products that each company designs, manufactures or uses, and how they make use of new technology.

Present your findings in the form of a brief word processed 'fact sheet' suitable for house buyers.

Activity 2.19

The use of continuously variable transmission systems (CVT) in passenger car applications is increasing with the development of suitable electronic hardware to achieve the ultimate objectives of increasing efficiency and reducing vehicle emissions. One of the transmission systems being investigated involves a split power configuration. This offers an improvement in efficiency as well as the ability to control the vehicle around the geared neutral state by means of torque control rather than ratio control which is commonly used in CVT systems based on the Van Doorne belt drive. Use information sources, such as your school or college library or the World Wide Web, to investigate the latest developments in CVT technology. Write a brief article for a car enthusiast magazine that presents your findings and highlights the advantages of this new technology.

Case study   The recordable compact disk

Conventional CD-ROMs, like audio compact disks, are made up of three basic layers. The main part of the disk consists of an injection-moulded polycarbonate *substrate* which inorporates a spiral track of *pits* and *lands* that are used to encode the data that is stored on the disk. Over the substrate is a thin aluminium (or gold) reflective layer, which in turn is covered by an outer protective lacquer coating.

Information is retrieved from the CD by focusing a low-power (0.5 mW) infrared (780 nm) laser beam onto the spiral track of pits and lands in the disk's substrate. The height difference between the pits and the adjacent lands creates a phase shift causing destructive interference in the reflected beam. The effect of the destructive interference and light scattering is that the intensity of the light returned to a photodiode detector is modulated by the digital data stored on the disk. This modulated signal is then processed, used for tracking, focus, and speed control, and then decoded and translated into usable data.

Conventional compact disks and CD-ROMs only support playback (or reading) of the data stored in them. In recent years new technology has appeared which supports both playback (reading) and recording (writing). This technology has resulted in two new types of compact disk; the CD-R and the CD-E (or CD-RW).

CD-R stands for *compact disk recordable*, a compact disk that supports recording as well as playback. With a conventional (non-recordable) CD, data can only be read from the disk – in other words, it cannot be written to. CD-R allows data to be read from the disk and recorded to it. However, unlike CD-E (*compact disk erasable*) or CD-RW (*compact disk rewritable*), data can be written only *once* to the disc – it cannot subsequently be erased and then rewritten (note that writing is only possible if there is enough unused space on the disk).

The idea of a rewritable compact disk is not particularly new. As far back as 1988, the Tandy Corporation in the United States claimed that their Magnetic Media Research Center in Santa Clara, California had developed an erasable compact disk which was fully compatible with existing audio CD and CD-ROM drives. Known as the THOR-CD (Tandy high–intensity optical recording compact disc) this new media was promoted as a very significant technological breakthrough.

Today's CD-R disks use a four-layer system consisting of a polycarbonate substrate, a sensitive organic dye layer, a gold or silver alloy reflective layer, and a protective lacquer coating. Unlike a conventional CD, the substrate of a CD-R disk does not contain a track of pits and lands. Instead, the substrate contains a slightly eccentric spiral pregroove to provide *absolute time in pregroove* (ATIP) information. When the disk is rotated at the correct speed under the focused laser beam, the reflected light returned to the photodetector from the pregroove generates a 22.05 kHz carrier frequency present in the current passing through the photodetector. This carrier frequency is compared with an accurate reference frequency source in order to ensure correct tracking, motor control, and focusing. The carrier is also frequency-modulated (± 1 kHz) to encode additional data such as time code information, a value for the recommended optimum recording power for use with the particular media employed, and a disk application code.

The most important difference between a CD-R disk and an ordinary CD is the inclusion of a sensitive dye layer that is designed to absorb light with a wavelength of 780 nm and thus accept the data that is to be recorded on the disk. Several different types of dye are used including stabilized cyanine, phthalocyanine-based material, or metalized azo. Instead of recording the data in the form of a pattern of pits and lands it is written into the dye layer using a relatively high power laser (4–8 mW at single writing speed, 8–10 mW at double speed, 10–12 mW at quadruple speed and 12–14 mW at sextuple speed). The pulses of power from the laser cause heating in the dye layer (with localized temperatures in the range 250°C to 400°C) that produce regions of decreased reflectivity.

In a conventional CD the physical height change of a moulded pit produces dispersion of light when the disk is being read. In the CD-R the changes in reflectivity of the dye mimic the light-scattering effects of moulded pits, modulating the intensity of the light beam returned to the photodiode detector and providing an end result that is close to that of a conventional compact disk.

A 63-minute disk (containing 540 MB of data) is spun at a real-time *constant linear velocity* (CLV) of 1.4 m/s while a 74-minute disk rotates rather more slowly at 1.2 m/s CLV, creating smaller marks and lands and thereby increasing the capacity to 650 MB.

CD-E disks (also known as CD-RW disks) were first announced in 1995. CD-E allows data not only to be written, but rewritten to the disk many times over. To accomplish this, CD-E media employs a six-layer design consisting of a polycarbonate substrate, a lower dielectric layer, a phase-change alloy recording layer, an upper dielectric layer, an aluminium alloy reflector, and a protective lacquer overcoat.

As with CD-R disks, CD-E media uses a polycarbonate substrate into which a spiral pregroove is moulded. In addition to the motor control and other information carried in the ATIP of a CD-R disk, a CD-E disk contains information specific to its format, including indicative write power values, recording speed ranges, recommended erase/write power ratios, and other critical data. As with CD-R disks, CD-E media can be rotated at different CLVs.

Erasability of data is made possible with CD-E technology by virtue of a recording layer consisting of a quaternary phase-change alloy composed of silver, indium, antimony, and tellurium. This operates on the principle of changing the phase of the recording material between a highly reflective crystalline state and a low reflective amorphous state. Unlike other phase-change storage technologies that 'burn bright' or write data by creating areas of increased reflectivity on light-scattering background, CD-E is a 'burn dark' technology that writes lower reflectivity marks on a reflective surface. This ensures that the resulting track of amorphous marks and crystalline spaces is as optically close as possible to the pattern of pits and lands on a conventional moulded CD. In its deposited state, phase-change material is amorphous, but CD-E media is supplied from the factory in a crystalline 'pre-initialized' state.

To write data to a CD-E disk, a relatively powerful (8–14 mW) 780 nm laser beam is focused in the pregroove and the laser energy is absorbed and heats the crystalline phase-change alloy to its melting point (600°C). When it cools, rather than recrystallizing, *it revitrifies* into the amorphous state to create the phase change. For rewriting the disk, CD-E is a *direct overwrite* (DOW) system, which means new data is recorded over existing information without first erasing what is already there. To erase/overwrite data, a lower power (4–8 mW) laser is focused on the amorphous portions of the recording layer, which raises it to its glass transition temperature (200°C) transforming it to a crystalline state.

During recording, the laser power is modulated according to an involved write strategy. Unlike CD-R, where the laser runs at its full recording power for the full interval of time required to form the desired length of a mark, CD-E repeatedly pulses its laser between full write power and a much lower bias power (same as that which is used to read the disk). The laser is pulsed to prevent the alloy from accumulating too much heat and creating disproportionately large amorphous marks.

Instead of using the physical height change of a plastic pit to disperse light as occurs with moulded CDs, the written amorphous marks in the phase-change imitate the light-scattering effects of moulded pits. The change modulates from 15% to 25% the intensity of the light beam returned to the photodiode detector, and provides a result close enough to a conventional compact disk to be read by normal CD-ROM drives with a few modifications.

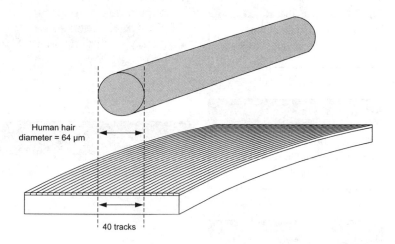

Figure 2.11 *Human hair compared with tracks on a CD*

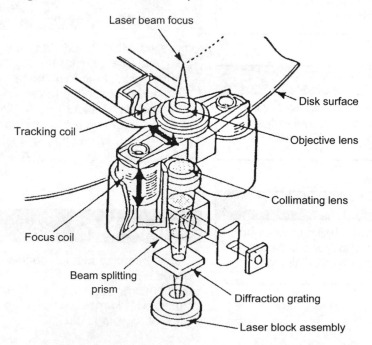

Figure 2.12 *Optical block assembly fitted to a CD-ROM drive*

Test your knowledge 2.27

What do the following abbreviations stand for?
(a) CD-R
(b) CD-E
(c) CD-RW
(d) CLV
(e) DOW
(f) ATIP.

Test your knowledge 2.28

Explain the function of the *pregroove* in a compact disk.

Test your knowledge 2.29

What are the requirements for the dye used in the recording layer of a CD-R disk?

Test your knowledge 2.30

Explain how data is recorded in a CD-R disk.

Test your knowledge 2.31

Explain what is meant by the term *direct overwrite* when applied to a CD-E disk.

Test your knowledge 2.32

Explain why it is necessary to control the power of the laser source in a CD-E drive.

Activity 2.20

CD-ROM standards are defined in a document known as the Orange Book. Find out about the Orange Book and the standards to which each part of it relates. Present your findings in the form of a brief word processed report.

Activity 2.21

Use the Internet or other information sources to search for manufacturers of recordable compact disks (both CD-R and CD-E). Obtain detailed specifications for each type of disk. Present your findings in the form of a set of word processed fact sheets.

Activity 2.22

Write a brief word processed article for a student newspaper explaining how CD-R and CD-E disks work.

Activity 2.23

Investigate the construction of a typical recordable CD drive (either internal or external) for use with a personal computer. Itemize each of the main components or sub-assemblies used and describe the materials and processes used in its manufacture. Present your findings in the form of a brief class presentation using appropriate visual aids. Comment on any safety issues that you encounter.

Activity 2.24

Obtain information and specifications on at least two DVD drives and explain why DVD drives are likely to replace conventional compact disk drives in the very near future. Write a word processed article for a student newspaper presenting your findings.

Review questions

1   For each of the engineering sectors listed below, identify a new product or service which uses new technology and, in each case, explain briefly how the technology in question has influenced the development of the product:

(a) Information and communications technology
(b) New materials and components
(c) Automation.

2   Explain the following terms used in computer aided engineering (CAE):

(a) computer aided design (CAD)
(b) computer aided manufacture (CAM)
(c) computer integrated manufacturing (CIM)
(d) computer numerical control (CNC).

3   Describe THREE different ways in which a computer can be used in the design and manufacture of a typical consumer electronics product.

4   Describe TWO applications of computer simulation in engineering. In each case explain why simulation is used.

5   Explain how an engineering company can use an Intranet to improve communications between employees and departments.

6   Explain how a search engine can be used to obtain information on polymer materials from the World Wide Web.

7   Explain TWO advantages of digital broadcasting compared with conventional analogue broadcasting techniques.

8   Explain the essential difference between thermoplastics and thermosetting materials (thermosets).

9   State TWO thermoplastic materials and TWO thermosetting materials.

10 Categorize each of the following materials as either a thermoplastic material or a themosetting material:

(a) Polythene
(b) Polyester
(c) Silicone
(d) Teflon.

11 State a typical application for each material in Question 10.

12 Explain what is meant by each of the following terms:

(a) microprocessor
(b) microcontroller
(c) microcomputer.

13 Explain, with the aid of a simple diagram, how a single-chip microcomputer is interfaced to external devices such as switches and relays.

14 State FOUR advantages of using programmable logic controllers (PLC) for controlling an industrial process.

15 State THREE types of I/O module commonly used in conjunction with a PLC.

16 Explain how data is stored on the surface of a compact disk.

17 Explain the differences between each of the following types of compact disk:

(a) a conventional compact disk (CD)
(b) a recordable compact disk (CD-R)
(c) an erasable compact disk (CD-E).

18 In relation to (c) in Question 15, explain how signals are recorded and erased.

19 Sketch a diagram showing the main parts of the optical block assembly used in a CD-ROM drive. Label your drawing clearly.

20 Apart from the signal information, what other information is encoded on the surface of a CD-R disk? Explain what this information is used for.

Unit 3 Engineering materials

Summary

All branches of engineering are concerned with the behaviour of materials. Civil Engineers need to ensure that the materials from which they build their roads, bridges and dams are suitable. Think about what might happen if they got their calculations wrong and selected the wrong type of material. For instance, if a bridge was built from the wrong material it might collapse under heavy loads. Even on a much smaller scale knowledge of materials is vitally important. An Electrical Engineer designing complex computer circuitry must ensure that the materials chosen do not prematurely deteriorate in service causing the computer to fail unexpectedly because, this too, could have catastrophic results. Imagine the consequences if the computer failed whilst controlling the automatic pilot of an aircraft just as the aircraft was landing! A study of materials is, therefore, an essential part of every engineer's vocabulary and for this reason alone, is worthy of study. However, I also hope that the more you learn about materials, the more you will come to realize what a fascinating and challenging a subject it is.

Materials and their properties

In the first element in this unit we are going to consider the important properties of the various types of materials used to make engineering products. Next we will look at typical measured values, for a variety of properties, for our chosen materials. Then, from appropriate literature and experimental tests, we will determine these data values and use them. Finally we will consider the way the chemical composition, types of bond, crystal and other structural features, affect the properties of the parent material.

Classes of material

What do you already know?

For most engineering applications the most important criteria for the selection of the material to be used are that the material does its job properly as cheaply as possible. Whether a material does its job depends on its *properties*, which are a measure of how it reacts to the various influences to which it is exposed. For example, loads, atmospheric environment, electromotive forces, heat, light, chemicals, and so on. To aid our understanding of the different types of materials and their properties I have divided them into four categories; metals, polymers, ceramics, and composites. Strictly speaking, composites are not really a separate group since they are made up from the other categories of material. However, because they display unique properties and are a very important engineering group, they are treated separately here.

Metals

You will be familiar with metals such as aluminium, iron, and copper, in the enormous variety of everyday applications: i.e. aluminium saucepans, copper water pipes and iron stoves. Metals can be mixed with other elements (often other metals) to form an *alloy*. Metal alloys are used to provide improved properties, i.e. alloys are often stronger or tougher than the parent pure metal. Other improvements can be made to metal alloys by *heat-treating* them as part of the manufacturing process. Thus steel is an alloy of iron and carbon and small quantities of other elements. If after alloying the steel is quickly cooled by quenching in oil or brine, a very hard steel can be produced. Much more will be said later about alloying and heat-treating metals.

Polymers

Polymers are characterized by their ability to be (initially at least) moulded into shape. They are chemical materials and often have long and unattractive chemical names. There is considerable incentive to seek more convenient names and abbreviations for everyday use. Thus you will be familiar with PVC (polyvinyl chloride) and PTFE (polytetraflouroethylene).

Polymers are made from molecules which join together to form long chains in a process known as *polymerization*. There are essentially three major types of polymer. *Thermoplastics*, which have the ability to be remoulded and reheated after manufacture. *Thermosetting plastics*, which once manufactured remain in their original moulded form and cannot be re-worked. *Elastomers or rubbers*, which often have very large elastic strains, elastic bands and car tyres are two familiar forms of rubber.

Ceramics

This class of material is again a chemical compound, formed from oxides such as silica (sand), sodium and calcium, as well as silicates such as clay. Glass is an example of a ceramic material, with its main constituent being silica. The oxides and silicates mentioned above have very high melting temperatures and on their own are very difficult to form. They are usually made more manageable by mixing them in powder form, with water, and then hardening them by heating. Ceramics include, brick, earthenware, pots, clay, glasses and refractory (furnace) materials. Ceramics are usually hard and brittle, good electrical and thermal insulators and have good resistance to chemical attack.

Composites

These combine the attractive properties of the other classes of material while avoiding some of their drawbacks. They are light stiff and strong, and they can be tough.

A composite is a material with two or more distinct constituents. So, for example, we can classify bricks made from mud and reinforced with straw which were used in ancient civilizations, as a composite. A versatile and familiar building material that is also a composite is concrete; concrete is a mixture of stones (aggregate), held together by cement. In the last 40 years there has been a rapid increase in the production of synthetic composites, those incorporating fine fibres held in various polymers.

Although not considered as a separate class of material, some *natural materials* exist in the form of a composite. The best known examples are wood, mollusc shells and bone. Wood is an interesting example of a natural fibre composite; the longitudinal hollow cells of wood are made up of layers of spirally wound cellulose fibres with varying spiral angle, bonded together with lignin during the growing of the tree.

Activity 3.1

How much do you already know about different materials and their properties? Test your knowledge by trying the exercise set out below. In attempting to tackle this task you might find it helpful to explore the objects that exist within your own home, and ask yourself why they are made from those particular materials. Complete the table by using a grading scheme such as: excellent, good, fair, poor; or high, above average, below average, low, or some other similar scheme.

Material properties	Classes of material			
	Metals	Polymers	Ceramics	Composites
Density				
Stiffness				
Strength				
Toughness				
Ductility				
Shock resistance				
Hardness				
Thermal conductivity				
Electrical conductivity				
Corrosion resistance				
Chemical stability				
Melting temperature				

The previous classification of materials is rather crude and many important sub-divisions exist within each category. The natural materials, except for those mentioned above under composites, have been deliberately left out. The study of materials such as wool and cotton is better placed in a course concerned with the textile industry. Here, we will be concentrating on the engineering application of materials and will only mention naturally occurring materials where appropriate.

What you may have discovered from Activity 3.1 was how difficult it is to make valid judgements, even using my rather crude classification of materials. You may also not have been entirely clear about the properties I listed. For instance, what do you understand by 'chemical stability'? How do we really define 'hardness'?

Properties of materials

To help you answer the questions resulting from Activity 3.1, we can now look at some of the more important properties of materials. These properties are often broken down into two major subdivisions; *mechanical properties* and *physical properties*. The latter includes; *electrical properties*, *magnetic properties, thermal properties* and *chemical properties*.

It can be argued that chemical properties are in themselves another major subdivision. To introduce the small amount of physical chemistry necessary to understand the environmental stability of materials, I will present the subject of corrosion and corrosion prevention separately.

We will be looking at all of the above properties but we will leave processing properties until we deal with material processing, where it fits more readily. Also, information on the cost of materials will be found later when we deal with the selection of materials for engineering products.

Mechanical properties

Mechanical properties are the behaviour of materials when subject to forces and include strength, stiffness, hardness, toughness, and ductility to name but a few. For the sake of completeness and precision most of the important mechanical properties are defined below. They make rather tedious reading, but are necessary to help you select appropriate materials for specific engineering functions.

When a material is subject to an external force, then the forces that hold the atoms of the material together (bonding forces), act like springs and oppose these external forces. These external forces may tend to stretch or squeeze the material or make two parts of the material slide over one another in opposite directions, by acting against the bonding forces (Figure 3.1).

When a material is subject to external forces that tend to stretch it, then it is said to be in *tension*. The ability of a material to withstand these tensile (pulling) forces is a measure of its *tensile strength* (Figure 3.2a). When a material is subject to forces that squeeze it then it is said to be in *compression*. The ability of a material to withstand these compressive forces is a measure of its *compressive strength* (Figure 3.2b). If a material is subject to opposing forces which are offset and cause one face to slide relative to an opposite face, then it is said to be in *shear* and the ability of the material to resist these shearing forces is a measure of its shear strength (Figure 3.2c).

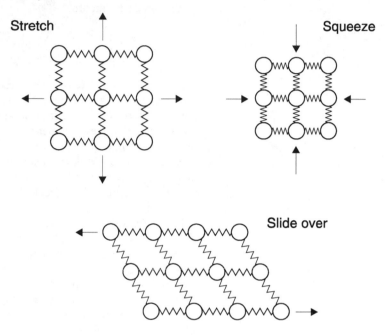

Figure 3.1 *Reaction of atomic bonds to external forces*

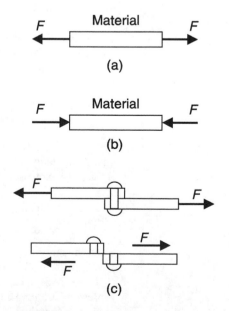

Figure 3.2 *(a) Tensile, (b) compressive and (c) shear*

In discussing the application of forces to materials, it is often desirable to be able to compare one material with another. For this reason we may not be concerned so much with the size of the force, as the force applied per unit area. Thus, for example, if we apply a tensile force F to a length of material over its cross-sectional area A, then the force applied per unit area is F/A. The term *stress* is used for the force per unit area.

$$\text{Stress} = \frac{\text{Force}}{\text{Area}}$$

The unit in which stress is measured is the Pascal (Pa), with 1 Pa being a force of 1 Newton per square metre, i.e. $1 \text{ Pa} = 1 \text{ N/m}^2$. Note that, in materials science, it is perhaps more convenient to measure stress in terms of Newtons per square millimetre (N/mm^2). This unit, moreover, produces a value that is easier to appreciate, whereas the force necessary to break (for example) a steel bar one square metre in cross-section is so large as to be difficult to visualize in ordinary measurable terms. The Greek letter (sigma σ) is often used to indicate stress.

The stress is said to be *direct stress* when the area being stressed is at right angles to the line of action of the external forces, as when the material is in tension or compression. Shear stresses are not direct stresses since the forces being applied are in the same plane (parallel with) the area being stressed.

The area used for calculating the stress is generally the original area that existed before the application of forces. This stress is often referred to as the *engineering stress*. The term *true stress* being used for the force divided by the actual area that exists while the material is in the stressed state.

Strain refers to the proportional change produced in a material as a result of the stress applied. It is measured as the number of millimetres of deformation (change in dimension) suffered per millimetre of original dimension and is a numerical ratio, therefore *strain has no units*.

When a material is subject to tensile or compressive forces and a change in length results then the material has been strained, this strain is defined as:

$$\text{Strain} = \frac{\text{change in length}}{\text{original length}}$$

For example, if we have a strain of 0.02. This would indicate that the change in length is $0.02 \times$ the original length. However, strain is frequently expressed as a percentage.

$$\text{Strain} = \frac{\text{change in length}}{\text{original length}} \times 100\%$$

Thus the strain of 0.02 as a percentage is 2%, i.e. here the change in length is 2% of the original length.

The Greek letter (epsilon ε) is the symbol normally used to represent strain. The symbol for length is (l) and the symbol for change in length is (δl), where δ is the Greek letter delta.

Example 3.1

A copper bar has a cross-sectional area of 75 mm² and is subject to a compressive force of 150 N. What is the compressive stress?

The compressive stress is the force divided by the area that is:

$$\text{compressive stress } (\sigma) = \frac{150}{75} \text{ N/mm}^2 = 2 \text{ N/mm}^2$$

$$\text{or compressive stress } (\sigma) = \frac{150}{75 \times 10^6} = 2 \text{ MN/m}^2 = 2 \text{ MPa}$$

Example 3.2

In a tensile test a specimen of length 140 mm, is subject to a tensile force that increases its length by 0.028 mm. What is the percentage strain?

The strain is the change in length divided by the original length that is

$$\text{strain } (\varepsilon) = \frac{0.028}{140} \times 100 = 0.02\%$$

We have already mentioned tensile, compressive and shear strength (Figure 3.2) but we did not give the general definition. The *strength* of a material is defined as *the ability of a material to resist the application of a force without fracturing*.

The stress required to cause the material to fracture *i.e. fracture stress* is a *measure of its strength* and requires careful consideration.

Hooke's Law

Hooke's Law states that *within the elastic limit of a material the change in shape is directly proportional to the applied force producing it.* What this means is that a linear (straight line) relationship exists between the applied force and the corresponding strain while the material is being elastically deformed (in other words, while it is still able to return to its original size when the external load (i.e. force) is removed).

A good example of the application of Hooke's law is the *spring*. A spring balance used for measuring weight force, for instance, where an increase in weight causes a corresponding extension (Figure 3.3).
If the load (weight) is increased sufficiently, there will come a point when the internal forces holding the spring material together start to break or permanently stretch and so, even after the load has been removed, the spring will remain permanently deformed, i.e. *plastically strained.*

By considering Hooke's law, it follows that *stress* is also directly proportional to strain while the material remains elastic, because stress is no more than force (load) per unit area. The stress required to first cause plastic strain is known as the *yield stress* (Figure 3.4).

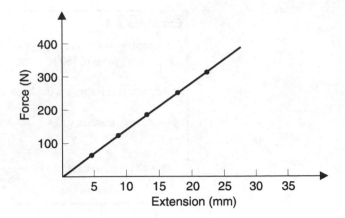

Figure 3.3 *Application of Hooke's Law to a spring*

For *metals*, the *fracture stress* (i.e. the measure of *strength*) is considered to be identical to the 0.2% yield stress. This is because, for all engineering purposes, metals are only used within their elastic range, so a strength measurement above the yield stress is of little value to engineers when deciding which metal to use. The *0.2% yield stress* is defined as the stress at which the stress–strain curve (Figure 3.4), for tensile loading, deviates by a strain of 0.2% from the linear elastic line. Note here that a permanent change in length (strain) has occurred, the material has been plastically deformed by 0.2% of its original length.

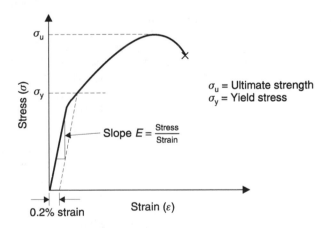

Figure 3.4 *The yield stress–strain curve for a metal, showing the measure of yield strength*

Where σ_f is the symbol for fracture stress (units MPa or MN/m^2) and σ_y is the yield stress (the stress needed to start to plastically deform or permanently strain the material), with the same units as fracture stress. For metals the fracture stress is the same in tension and compression.

For *polymers* the fracture stress is identified as the yield stress at which the stress–strain curve becomes markedly non-linear i.e., typically at strains of around 1%. Polymers are a little stronger (approximately 20%) in compression than in tension (Figure 3.5).

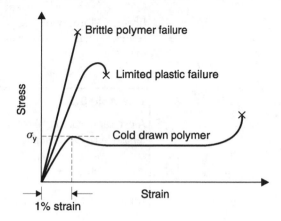

Figure 3.5 *Stress–strain curves for a polymer showing the 1% measure of yield strength*

For ceramics and glasses the strength depends strongly on the mode of loading. In tension, strength means the fracture strength given by the *tensile fracture stress* (symbol σ_f^t). In compression it means the crushing strength, given by the *compressive fracture stress* (symbol σ_f^c) which is much larger than the tensile fracture stress, typically 15 times as large. The tensile and compressive fracture stresses for a typical ceramic are shown in Figure 3.6.

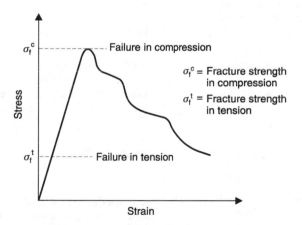

Figure 3.6 *The stress–strain curve for a ceramic showing the large variation in yield strength in tension (low) and compression (high)*

The symbol σ_u is used to indicate the *ultimate tensile strength*, measured by the nominal stress at which a bar of material loaded in tension separates (breaks). For brittle solids – ceramics, glasses and brittle polymers – it is the same as the fracture stress in tension (Figures 3.5 and 3.6). For metals, ductile polymers and most composites, it is larger than the fracture stress (σ_f) by a factor of between 1.1 and 3 because of work hardening (Figure 3.4), or in the case of composites, because of load transfer to the strong reinforcing fibres. More will be said about work hardening and fibre reinforcement of composites later when we deal with the structure of materials.

Example 3.3

A circular metal rod 10 mm in diameter has a yield stress of 210 MPa. What tensile force is required to cause yielding?

Since stress (s) $= \dfrac{\text{force}}{\text{area}}$

we must first calculate the cross-sectional area of the metal rod. Thus the area A is given by:

$A = \pi\, r^2 = \pi \times 5^2 = 78.54 \text{ mm}^2$

Therefore yield force required = yield stress × area
$$= 210 \times 10^6 \times 78.54 \times 10^{-6} = 16{,}493 \text{ N}$$

Young's modulus

Stiffness can be defined as the ability of a material to resist deflection when loaded. For example, a material's resistance to bending is a measure of its stiffness. When a material is subject to external bending forces the less it gives, the stiffer it is. Thus stiffness is related to the stress imposed on the material and the amount of movement or strain caused by this stress. We mentioned earlier the linear relationship between stress and strain on the stress–strain graph. A measure of the stiffness of a material when strained in tension can be obtained from the graph by measuring the slope of the linear part of the graph (see Figure 3.4). This slope (stress)/(strain) is known as the *elastic modulus* or *Young's modulus of elasticity*.

$$\text{Modulus of elasticity} = \frac{\text{stress}}{\text{strain}}$$

The modulus of elasticity has units of stress (often quoted as GPa), since strain has no units. The symbol for the elastic modulus is E. Figure 3.4 indicates the relationship between, stress, strain and the elastic modulus.

In addition to the elastic modulus E, when dealing with three-dimensional solids, there are two other moduli that are worthy of mention. The *shear modulus, G,* describes the rigidity of a material subject to shear loading; and the *bulk modulus, K,* describes the effect of a material when subjected to external pressure.

So far I have given a fairly rigorous definition of the strength of a material, as well as defining stress, strain and material stiffness. You have already met quite a lot of terms with which you may not be completely familiar. To aid your understanding and to refresh your memory, it is time to try the problems and the activity that follow.

Test your knowledge 3.1

Define as accurately as you can the terms listed below and, for each, give (if appropriate) the units and the normal symbol used to identify them:

- engineering stress
- true stress
- tensile strain
- strength
- ultimate strength
- modulus of elasticity.

Test your knowledge 3.2

Explain briefly how the fracture strength and yield stress are related with respect to metals, polymers and ceramics. Clearly defining fracture strength and yield stress, as part of your explanation.

Test your knowledge 3.3

Given that the moduli, *E*, *G* and *K* are related in the following ways:

$$E = \frac{3G}{\left(1 - \frac{G}{3}K\right)}$$

$$G = \frac{E}{2(1 + 2v)}$$

$$K = \frac{E}{3(1 - 2v)}$$

determine the shear modulus and bulk modulus for a steel which has an elastic modulus of 210 GN/m² and Poisson's ratio *v* is 0.3.

Activity 3.2

The abbreviations for a number of well known polymers and elastomers are given below. Find, from appropriate literature, the full chemical name and write this alongside each abbreviation.

Abbreviation	*Full chemical name*
LDPE	
PP	
PC	
PMMA	
PTFE	
PF	
UF	
PU	
SBR	
PB	

More mechanical properties

Let us now consider a few more mechanical properties, which are presented here with examples of materials that display these properties.

Ductility, the ability to be drawn into threads or wire. Examples include; wrought iron, low carbon steels, copper, brass and titanium.

Brittleness, the tendency to break easily or suddenly with no prior extension. Examples include; cast iron, high carbon steels, brittle polymers, concrete, and ceramic materials.

Malleability, the ability to plastically deform and shape a material by forging, rolling or by the application of pressure. Examples include; gold, copper and lead.

Elasticity, the ability of a material to deform under load and return to its original shape once the external loads have been removed. Internal atomic binding forces are only stretched, not broken and act like minute springs to return the material to normal, once the force has been removed. Examples include; rubber, mild steel and some plastics.

Plasticity is the readiness to deform to a stretched state when a load is applied. The *plastic deformation is permanent* even after the load has been removed. Plasticine exhibits plastic deformation.

Hardness, the ability to withstand scratching (abrasion) or indentation by another body. It is an indication of the wear resistance of a material. Examples of hard materials include; diamond high carbon steel and other materials that have undergone a hardening process.

Fatigue is a phenomenon by which a material can fail at much lower stress levels than normal when subjected to cyclic loading. Failure is generally initiated from micro-cracks on the surface of the material.

Creep may be defined as the time dependent deformation of a material under load, accelerated by increase in temperature. It is an important consideration where materials are subjected to high temperatures for sustained periods of time, e.g. gas turbine engine blades.

Toughness, in its simplest form, is defined as the ability of a material to withstand sudden loading. It is measured by the total energy that a material can absorb, that just causes fracture. Toughness (symbol, G) must not be confused with strength that is measured in terms of the stress required to break a standard test piece. Toughness has the units of energy per unit area, i.e. MN/m^2.

Fracture toughness (symbol K_c) measures the resistance of a material to the propagation (growth) of a crack. Fracture toughness for a material is established by loading a sample containing a deliberately introduced crack of length $2c$ (Figure 3.7), recording the tensile stress (σ_c) at which the crack propagates.

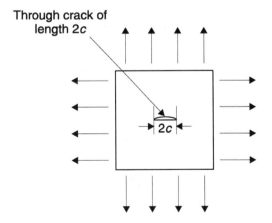

Figure 3.7 *Establishment of fracture toughness K_c by loading a sample with a deliberately introduced through-crack of length 2c and recording tensile stress*

The quantity K_c is then calculated from:

$$K_c = Y\sigma_c (\pi c)^{1/2}$$

normal units for K_c being MPam$^{1/2}$ or MN/m$^{3/2}$

The fracture toughness (K_c) is related to the toughness by the relationship:

$$G_c = \frac{K_c^2}{E\left(1 - v^2\right)}$$

Y is a geometric factor, near unity, which depends on the geometry of the sample under test, v is known as Poisson's ratio where:

$$\text{Poisson's ratio } (v) = \frac{\text{lateral strain}}{\text{axial strain}} \text{ of a strained material under load}$$

Note: axial strain is strain along the longitudinal axis of the material.

Physical properties

In the previous section we dealt in some detail with the mechanical properties of materials, it is now time to concentrate on their physical properties, these are subdivided into *electrical* and *magnetic, thermal, density* and *optical*. With the exception of a brief statement on photoconduction, optical properties will not be considered in this course. We will start by briefly discussing some of the electrical and magnetic properties of materials including *electrical conductivity, superconductivity, semiconductors, magnetization,* and *dielectrics*.

In many applications, the electrical properties of a material are of primary importance. Copper wire is chosen for electrical wiring because of its extremely high electrical conductivity. You may also remember that copper is also a very ductile material and so is easily drawn or extruded into shape. These two properties make copper ideal for electrical wiring.

Conductors are materials having outer electrons that are loosely connected to the nucleus of their atoms and can easily move through the material from one atom to another. *Insulators* are materials whose electrons are held firmly to the nucleus.

For a conducting material the electromotive force (e.m.f.) measured in volts, V, the current measured in amperes, I, and the resistance measured in Ohms, R, are related by Ohm's law. This states that:

$$V = I R$$

The resistance to current flow in a circuit is proportional to the length, l, and inversely proportional to the cross sectional area, A, of the component. The resistance can be defined as:

$$R = \frac{\rho l}{A} \text{ where } \rho \text{ is the electrical } \textit{resistivity}$$

or

$$R = \frac{l}{\sigma A} \text{ where } \sigma \text{ is the electrical } \textit{conductivity}$$

The unit of *resistivity* is the ohm metre (Ωm), and the unit of *conductivity* is the Siemen per metre (S/m). It can be seen that the electrical conductivity is the inverse of the electrical resistivity. Both of these properties, conductivity and resistivity, are inherent in each material. Typical tables of values normally show only the resistivity of a material. Very high values suggest very good resistive characteristics and, conversely very low values indicate good conductivity.

Metals which have many free electrons are very good electrical conductors, their *conductivity* is reduced with increase in temperature. The rate of increase of electrical resistivity is given by the temperature resistivity coefficient. The resistivity for some of the more common materials together with their temperature resistivity coefficients are given in Table 3.1.

Table 3.1 *Electrical resistivity and temperature resistivity coefficients for some of the more common materials*

Material	Electrical resistivity at 20° C (Ωm)	Temperature resistivity coefficient (10^{-3}/K)
Aluminium	27×10^{-9}	4.2
Brass	69×10^{-9}	1.6
Constantan	490×10^{-9}	0.02
Copper	17×10^{-9}	4.3
Duralumin	50×10^{-9}	2.3
Gold	23×10^{-9}	3.9
Lead	206×10^{-9}	4.3
Mild steel	120×10^{-9}	3.0
Polythene	100×10^{9}	–
Rubber	Approx. 10×10^{9}	–
Silver	16×10^{-9}	4.1
Tungsten	55×10^{-9}	4.6

Superconductivity – superconducting materials are those where the resistance falls to zero when they are cooled to a critical temperature and any magnetic fields are minimized. At these temperatures the superconducting material offers no resistance to the flow of current and so there are no wastages due to heat generation. All pure crystals would act as superconductors at zero degrees Kelvin, the problem of course is to maintain and use them at this temperature!

Serious research into raising the temperature at which selected materials will superconduct has taken place over the past two decades. By 1986 the critical temperature of a series of ceramic copper oxides had been raised to 77 K. This enabled liquid nitrogen to be used as the cooling medium, rather than liquid helium, the latter being expensive and providing only limited application of these materials. Since this date other compounds based on non rare-earth metal oxides have exhibited temperatures up to about 125 K. There is still a little way to go before we arrive at high temperature superconductors. As superconduction temperatures rise, then more and more industrial applications become feasible. Even today it seems possible that the new materials that are constantly being discovered, could be used to advantage in the electrical power generation industry. Other applications might include the rail transport industry, computing and electronics.

We have already discovered the fact that for metals and most other conductors, the conductivity decreases with increase in temperature. The resistance of insulators remains approximately constant with increase in temperature. *Semiconductors* behave in the opposite way to metals and as the temperature increases the conductivity of a semiconductor increases.

The most important semiconductor materials are *silicon* and *germanium,* both are used commonly in the electronics industry. As the temperature of these materials is raised above room temperature, the resistivity is reduced and ultimately a point is reached where effectively they become conductors. This increase in conductivity with temperature has made semiconductors an essential part of thermistors, where they can be used to sense the temperature and activate a signal when a predetermined temperature is reached. Other uses for semiconductors include, pressure transducers (energy converters), transistors and diodes. The effect of temperature on conductors, insulators and semiconductors is illustrated diagrammatically in Figure 3.8. Note that for a specimen of each of these materials it is assumed that they start with the same resistance at say room temperature (20°C). This, of course, implies that each specimen will have completely different physical dimensions (see definition of resistance).

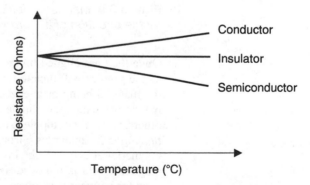

Figure 3.8 *The effect of temperature increase on conductors, insulators and semiconductors*

In order to control the conductivity of semiconductors, small amounts of impurity atoms are introduced and this is called *doping*. If antimony, arsenic or phosphorus are used as a doping agent then an n-type semiconductor is formed, since these doping agents add electrons to the parent semiconductor thus increasing its negative (n) charge. Conversely if gallium, indium or boron are added to the parent material then a p-type semiconductor is formed, with effectively an increased number of absences of electrons or *holes* which act like positive (p) charges.

In *photoconduction* a beam of light can be directed onto a semiconductor positioned in an electrical circuit. This can produce an electric current caused by the increased movement of electrons or holes in the atoms of the semiconductor material. This property is used in photoelectric components such as solar cells where the photoconductive property of the materials is used to produce power from the rays of light produced by the Sun. Other uses include electronic eyes that trigger the power to open garage doors, or automatic lighting where the photoelectric cell is activated by fading daylight.

Magnetization   All substances are magnetized under the effect of a magnet field. A magnet dipole (an atom which has its own minute magnetic field) is formed by the rotation of each electron about its own axis (electron spin) and also its rotation about the nucleus of the atom (Figure 3.9). This induces a magnetic field in the material that is increased in strength when the dipoles are aligned.

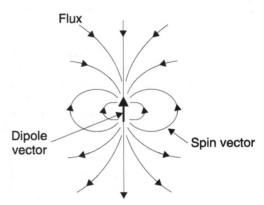

Figure 3.9 *Magnetic dipole moment of an electron (indicated by the direction of the arrow)*

Once they are aligned by an external magnetic field, the orientation of the dipoles will dictate what type of magnetism is created within the material being magnetized. Some materials called *paramagnetic*, magnetize with their dipole axis in the direction of the field and others, with their dipole axis perpendicular to the field these are *diamagnetic* materials. Iron, which shows a very pronounced magnetic effect and retains some residual magnetism when the magnetic field is removed, is *ferromagnetic*. Such materials can be used for permanent magnets.

If a bar of ferromagnetic material is wrapped with a conductor carrying a direct current, the bar will develop a north pole at one end and a south pole at the other (Figure 3.10).

The strength of the poles developed depends upon the material used and the magnetizing force. The magnetizing force is the result of the number of turns of conductor per unit length of the bar and the number of turns carried. The magnetic field produced as a result of the current and conductor length is denoted by the symbol, H, and is measured in amperes per metre (A/m). The magnetizing force is

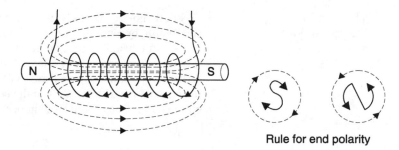

Rule for end polarity

Figure 3.10 *Magnetic flux developed from the current-carrying conductor wound around the ferromagnetic material*

increased, decreased, or reversed by respectively increasing, decreasing, or reversing the current. The strength of the magnet so produced is called the *magnetic flux density,* which indicates the degree of magnetization that can be obtained. The magnetic flux density is denoted by the symbol, *B,* and is measured in Tesla (T); it is related to the applied field, *H*, by:

$$B = \mu H$$

where μ is the magnetic *permeability* and is measured in Henry/metre (H/m). The ratio of this magnetic permeability (m) to the permeability of free space (μ_0) measured in a vacuum, gives an indication of the degree of magnification of the magnetic field and is known as the relative permeability (μ_r)

Magnetization and demagnetization of magnetic materials results in producing a variable magnetic flux density, *B*, as a result of the magnetic field strength, *H*. A plot of *B* versus *H* results in a diagram known as a *hysteresis loop* (Figure 3.11). The area under the curve represents irreversibly lost electromagnetic energy which is converted to heat.

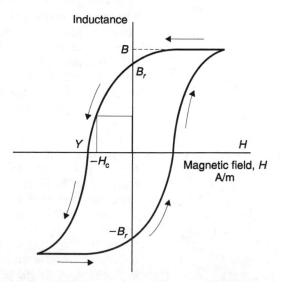

Figure 3.11 *Typical hysteresis curve for a ferromagnetic material. The area under the curve indicates the lost electromagnetic energy dissipated as heat*

Soft magnetic materials (temporary magnets) have a relatively small loop, because they are easily magnetized and demagnetized and so little heat is generated during the process. Hard magnetic (permanent magnets) have a large loop, due to the fact that they find difficulty in dissipating their residual magnetism. Thus *remanence* (or residual magnetism, B_r) and the *coercive force*, H_c, necessary to eliminate remanence are large in permanent magnets (Figure 3.12).

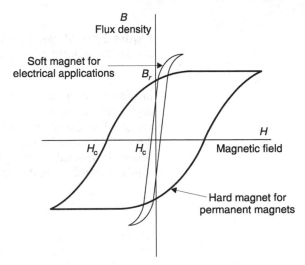

Figure 3.12 *Typical hysteresis curve for hard and soft magnetic materials*

Temporary magnets are soft iron or ferrous alloys; permanent magnets are hard steels, hard alloys, or metal oxides. Temporary magnets are used for alternating current applications such as a.c. motors, electromagnets and transformers.

A *dielectric* is an insulating medium separating charged surfaces. The most generally useful electric property of nonmetals is their high electrical resistance. Probably, the most common applications of ceramics, apart from their structural uses, depend upon their insulating properties. Breakdown of electrical insulators occurs either along the surface or through the body of the insulator. Surface breakdown is promoted by moisture and other surface contaminants. Water absorption can be minimized by applying a glaze to the surface of a ceramic.

Volume breakdown occurs at voltages high enough to accelerate individual free electrons to energies that will break the bonds holding the atoms of the material together, this causes a large number of electrons to break free at once and become charge carriers. The dielectric strength is the breakdown voltage for volume breakdown, it is measured in units of volts/m. The dielectric constant is the constant of proportionality between the charge stored between two plates of a capacitor separated by a dielectric, when compared with the charge stored when the plates are separated by a vacuum.

The ratio of the charge density, σ, to the electric field strength, E, is called absolute permittivity, ε, of a dielectric. It is measured in Farads/metre, F/m. The permittivity of free space measured in a vacuum is a constant, given by:

$$\varepsilon_0 = 8.55 \times 10^{-12} \text{ F/m}$$

Relative permittivity is a ratio given by:

$$\varepsilon_r = \frac{\text{flux density of the field in the dielectric}}{\text{flux density of the field in a vacuum}}$$

Let us now continue with our discussion on the physical properties of materials by considering their *density*.

The *density*, ρ, of a substance is its mass per unit volume. The basic SI units of mass and volume are, respectively, the kilogramme and the metre cubed, so the basic units of density are kilogramme per cubic metre (kg/m^3).

The *specific weight*, w, of a substance is its weight per unit volume. Since the weight of a substance is mass multiplied by acceleration due to gravity, g, the relationship between specific weight and density is:

specific weight = density $\times g$

$$w = \rho g$$

Density is a property of the material that may be subject to small changes by mixing one material with another. Significant changes in density are only accomplished with composite materials where large percentages of each material can be altered. The density of a composite can be calculated from the proportions of the constituent materials. For example, consider a composite that consists by volume of 60% epoxy resin matrix material to which is added 40% by volume carbon fibre. If we assume that the carbon fibre has a density of 1500 kg/m^3 and the epoxy resin has a density of 1100 kg/m^3. Then the density of the composite is given by:

$$\rho_c = 0.6\rho_m + 0.4\,\rho_f = (0.6 \times 1800) + (0.4 \times 1400) = 1640 \text{ kg/m}^3$$

where ρ_c = density of the composite, ρ_m = density of the matrix material and ρ_f = the density of the reinforcing carbon fibre.

The above equation is often referred to as an *equation of mixtures*, it has many uses particularly in the study of composite materials.

Activity 3.3

A small engineering company, Thames Magnetic Components, has commissioned you to carry out some research into materials suitable for use in the cores of a new range of large industrial transformers that they intend to manufacture. They have asked you to summarize the essential electromagnetic properties of a suitable core material and explain how these properties differ from those

Test your knowledge 3.4

What properties must dielectric materials used in the construction of capacitors possess? Name three materials that possess these properties.

Test your knowledge 3.5

Using the appropriate formula and Table 3.1, determine the resistance of a 50 m length of copper cable having a cross-sectional area of 1.5 mm².

used in the range of permanent magnets that make up their current product range. Write a briefing paper for the Technical Director, present your report in word processed form and include relevant diagrams and technical specifications.

Activity 3.4

An aerospace company, Archer Avionics, are about to manufacture an electronic control system that will be fitted to the propulsion unit of a deep space probe. This device is expected to be subject to extreme variations in temperature (e.g. from –50° to +150°C). Prepare a presentation (using appropriate visual aids) for the Board of Directors of Archer Avionics, explaining the effect of such a wide variation in temperature on the behaviour of the conductors, insulators and semiconductors that will make up the electronic system. Your presentation should last no more than 10 minutes and should include relevant graphs and material specifications. You should allow a further 5 minutes for questions.

Thermal properties

Let us now turn our attention to the *Thermal Properties* of materials. The way heat is transferred and absorbed by materials is of prime importance to the designer. For instance the walls of a house need to be constructed from materials that retain heat in winter and yet prevent the house from over heating in the summer. In a domestic refrigerator, the aim is to prevent the contents from absorbing heat. The thermal expansion of materials and their ability to withstand extremes of temperature are other characteristics that must be understood.

We therefore require knowledge of the most important thermal properties of materials, *specific heat capacity, thermal conductivity and thermal expansion*, as well as information on freezing, boiling and melting temperatures.

The specific heat capacity of a material is the quantity of heat energy required to raise the temperature of 1 kg of the material by one degree. The symbol used for specific heat capacity is c and the units are (J/kg °C) or (J/kg °K). If we just consider the units of specific heat capacity we can express it in words as:

$$\text{specific heat capacity} = \frac{\text{heat energy supplied}}{\text{mass} \times \text{temperature rise}}$$

Some typical values for specific heat capacity are given in Table 3.2.

Table 3.2 *Typical values for specific heat capacity for some common materials*

Material	Specific heat capacity (c) at 0°C (J/kgK)
Aluminium	880
Copper	380
Brass (65% Cu–35% Zn)	370
Iron	437
Lead	126
Mild steel	450
Silver	232
Tin	140
Brick	800
Concrete	1100
Polystyrene	1300
Porcelain	1100
Rubber	900

Thermal expansion takes place when heat is applied to a material and the energy of the atoms within the material increases, which causes them to vibrate more vigorously and so increase the volume of the material. Conversely, if heat energy is removed from a material contraction occurs in all directions.

The amount by which unit length of a material expands when the temperature is raised by one degree is called the *coefficient of linear expansion* of the material and is represented by the Greek letter (α) alpha. The units of the coefficient of linear expansion are usually quoted as just /K or K^{-1}.

Typical values for the coefficient of linear expansion of some materials are given in Table 3.3.

We can use the value of the coefficient of linear expansion of materials to determine changes in their length, as the temperature rises or falls. If a material, has initial length l_1 at a temperature t_1 and has a coefficient of linear expansion α, then if the temperature is increased to t_2, the new length l_2 of the material is given by:

new length = original length + expansion, i.e.

$$l_2 = l_1 + l_1 \alpha (t_2 - t_1)$$

or since $t_2 - t_1$ is often expressed as Δt (change in temperature) we have new length $l_2 = l_1 + l_1 \alpha \Delta t$

Table 3.3 *Typical values for coefficient of linear thermal expansion for some common materials*

Material	Coefficient of linear thermal Expansion (α) $(10^{-6}/K)$
Aluminium	23
Copper	16.7
Brass (65% Cu–35% Zn)	18.5
Invar	1
Iron	12
Lead	29
Magnesium	25
Mild steel	11
Silver	19
Tin	6
Brick	3–9
Concrete	11
Graphite	2
Polyethylene	300
Polyurethane foam	90
Polystyrene	60–80
Porcelain	2.2
PVC (plasticized)	50–250
Pyrex glass	3
Rubber	670

Example 3.4

In a domestic central heating system a 6 m length of copper pipe contains water at 8°C when the system is off, the temperature of the water in the pipe rises to 65°C when the system is in use. Calculate the linear expansion of the copper pipe.

From Table 3.3 we see that the coefficient of linear expansion for copper is 16.7 x 10^{-6}/°K, then using:

linear expansion of pipe = $l_1 \alpha t$ = 6 x 16.7 x 10^{-6} x (65 − 8)
= 0.0057 m

we see that the pipe expands by 5.7 mm over the 6 m length.

Thermal conductivity

Conduction is the transfer of energy from faster more energetic molecules to slower adjacent molecules, by direct contact. The ability of a material to conduct heat is measured by its *thermal conductivity, k.* The units of thermal conductivity are Watts per metre Kelvin (W/mK or $Wm^{-1}K^{-1}$). Typical values for some common materials are shown in Table 3.4.

Table 3.4 *Typical values for thermal conductivity for some common materials*

Material	Thermal conductivity (k) at 0°C (W/m K)
Aluminium	235
Copper	283
Brass (65% Cu–35% Zn)	120
Invar	11
Iron	76
Lead	35
Magnesium	150
Mild steel	55
Silver	418
Tin	60
Brick	0.4–0.8
Concrete	10.1
Graphite	150
Polyethylene	0.3
Polyurethane foam	0.05
Polystyrene	0.08–0.2
Porcelain	0.8–1.85
PVC (plasticized)	0.16–0.19
Pyrex glass	1.2
Rubber	0.15

A metal when left in a cold environment quickly feels cold to the touch and when brought into contact with heat quickly feels hot to the touch. Generally *metals* are *good conductors* of heat. Often if a material is a good thermal conductor, it is also a good electrical conductor, silver, copper and aluminium are all good conductors of both heat and electricity. Table 3.4 shows the low values of thermal conductivity for brick, porcelain, PVC and rubber, these are all good *insulators*. Air is also an excellent insulator, double glazing requires an airgap between the external and internal panes of glass, this provides good insulation from the cold and also helps prevent condensation forming between the panes. Expanded foams also use the properties of air to provide good insulation, foam cavity wall insulation for example.

Corrosion and the deterioration of materials in service

Corrosion may be defined as a chemical process, where metals are converted back to the oxides, salts and other compounds from which they were first formed. The corrosion of metals is therefore a natural process and in trying to combat it, we are wrestling with the forces of nature! The chemical stability of materials (in particular metals) has long been the subject of much research, since the consequences of premature failure of materials by corrosive influences can be disastrous.

Corrosive attack frequently occurs in combination with other mechanisms of failure, such as corrosion-fatigue, erosion and stress corrosion. Many environmental factors help to promote corrosion including; moist air and industrial pollutants such as dirt, acids, dirty water and salts. Corrosion may occur at elevated temperatures in materials that at lower temperatures are inert.

Corrosion then is the chemical means by which metallic materials deteriorate and fail. Two basic mechanisms have been recognized: *direct chemical attack* and *electrochemical attack.*

Direct chemical attack results in a uniform reaction over the entire exposed surface. Usually, a scale or deposit of uniform thickness is produced on the metallic material. This deposit may adhere (stick) to the surface or remain as loose flakes, the rusting of an iron bar left in the open is an example of the later. An example of a material that is subject to direct chemical attack, where the products of corrosion (the corrosive oxides) adhere strongly to the surface of the metal, is aluminium. When this happens the oxide layer formed protects the metal underneath from further attack by adhering firmly to it. This process is known as *passivation* and we say that aluminium is a *passive* metal.

Electrochemical attack is characterized by the establishment of an electrochemical cell, this is formed when two metals in electrical contact are placed in a conducting liquid, the *electrolyte*. The cell permits electroplating or corrosion, depending on the source of electrical potential.

In *electroplating* the electro-chemical cell, consists of two electrodes (which may or may not be of the same material), the electrolyte and an *external* electric source such as a battery. It is found that an oxidation reaction takes place at electrode A (Figure 3.13).

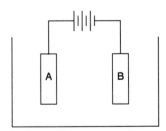

Figure 3.13 *In electroplating (external energy source) it is found that oxidation reaction takes place at electrode A*

An *oxidation reaction increases the energy of the atom.* In this case by forming an *ion* (an atom with an absence or excess of electrons) and a free electron. This may be represented by using a simple chemical equation:

$$M \rightarrow M^+ + e^-$$

What we are saying is that the metal (M) has been subject to an oxidation reaction and form a metal ion (M^+), in this case an atom that has lost *one* electron (e^-). The electrode, at which an oxidation reaction takes place is defined as the *anode*. Note that, *corrosion always takes place at the anode*. Then at electrode (A), the ions go into the electrolyte and the electrons go into the external circuit. At electrode (B) a *reduction reaction* takes place. In this case, the ions in the electrolyte combine with electrons from the external circuit and form atoms which are deposited on the electrode. This can again be illustrated using a chemical formula, i.e.

$$M^+ + e^- \rightarrow M$$

The electrode, at which the reduction reaction takes place is known as the *cathode*.

For electroplating the battery, provides the electrical potential to move the electrons, giving rise to an electric current in the external circuit. In practice, corrosion occurs in the absence of a battery. This can be demonstrated by a *galvanic cell*. A galvanic cell may be formed by two dissimilar metals in electrical contact immersed in an electrolyte, with no battery. Figure 3.14 shows a typical arrangement for a zinc-copper galvanic cell. The chemical symbol for zinc is (Zn) and for copper (Cu).

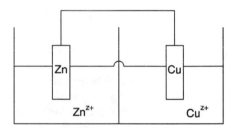

Figure 3.14 *In the copper–zinc galvanic cell an oxidation reaction takes place at the zinc electrode and it preferentially corrodes*

It is found that an oxidation reaction takes place at the Zn electrode. This electrode is thus the anode and it corrodes, i.e.

$$Zn \rightarrow Zn^{2+} + 2e^-$$

At the copper electrode, the cathode, a reduction reaction occurs, i.e.

$$Cu^{2+} + 2e^- \rightarrow Cu$$

The electrons in the external circuit flow between the two electrodes and so there must be an electrical potential difference between them.

Electroplating rarely occurs during electrochemical corrosion. The reduction reaction tends to form either a gas, liquid or solid by-product at the cathode.

A cathodic reaction is a reduction reaction, some possible cathodic reactions are given below.

1. Electroplating $M^+ + e^- \rightarrow M$

2. The hydrogen electrode – here hydrogen gas is liberated at the cathode:

$$2H^+ + 2e^- \rightarrow H_2$$

3. Water decomposition – here water is being broken down by the addition of electrons from the external circuit, to form hydrogen gas and hydroxyl (hydrogen–oxygen) ions

$$2H_2O + 2e^- \rightarrow H_2 + 2(OH)^-$$

Note that *chemical equations balance*. Taking the above equation, for example, two molecules of water plus two electrons form two atoms of hydrogen, diatomically bonded (H_2) and two hydroxyl ions.

4. The oxygen electrode – here oxygen combines with water and the external electrons to form hydroxyl ions.

$$O_2 + 2H_2O + 4e^- \rightarrow 4(OH)^-$$

5. The water electrode – in this case water is formed as a product.

$$O_2 + 4H^+ + 4e^- \rightarrow 2H_2O$$

When hydroxyl ions (OH^-) form as a result of a reaction, they can combine with other available ions to form a solid or sediment, e.g. Fe $(OH)_3$ which is rust.

Note that in all chemical equations the constituents that combine to make up the product, *the reactants*, are always placed at the tail of the arrow, the arrow indicates the direction of the reaction and *the products* (the result of the chemical reaction) are always placed after the head of the arrow.

If the zinc electrode, in the zinc–copper cell, is immersed in a 1 molar* electrolyte and the copper electrode (both at 25°C) is immersed in a 1 molar electrolyte, then a potential difference between the electrodes of 1.1 V occurs. The zinc electrode is at a lower potential than the copper electrode and so electrons flow, via the external circuit, from the zinc to the copper electrode.

The electrode potential of a metal is related to its ability to produce free electrons. As with all potential difference (PD) measurements, it is necessary to have a reference. In this case the PD

---

* A one molar solution of electrolyte is a special measure of the strength of the electrolyte, which is easily reproducible, where the electrolyte contains 1 g atomic mass of zinc ions, i.e. 6×10^{23} ions.

of a metal is measured with respect to the *hydrogen half-cell*, i.e. zero potential is assigned to the oxidation reaction:

$$H_2 \rightarrow 2H^+ + 2e^-$$

The hydrogen half-cell consists of a platinum electrode immersed in a 1 molar solution of hydrogen ions through which hydrogen gas is bubbled. The other half of the galvanic cell consists of a metal electrode immersed in a 1 molar solution of its own ions. Both half cells are at 25°C. For example, if zinc is compared with the hydrogen half-cell under the conditions mentioned above, its potential difference is −0.76V. This implies that the electrons flow via the theoretical circuit, from the zinc electrode to the hydrogen electrode. Zinc is the anode and corrodes, the hydrogen electrode is the cathode.

In the case of silver (Ag), the potential difference, or electrode potential is found to be +0.8V. Thus silver becomes the cathode and corrosion does not occur. The electrode potential of metals have all been measured in this way and a table of values produced. This table is known as the *electrochemical* or *redox series*. Some of the more common metals are listed in Table 3.5.

Table 3.5 *The electrochemical (redox) series for some elements*

Element	Potential E (V)
Lithium	−3.05
Potassium	−2.93
Caesium	−2.92
Calcium	−2.87
Sodium	−2.71
Magnesium	−2.37
Aluminium	−1.66
Titanium	−1.63
Zinc	−0.76
Chromium	−0.74
Iron	−0.44
Cadmium	−0.40
Nickel	−0.25
Tin	−0.14
Lead	−0.13
Hydrogen	0.00
Copper	+0.34
Silver	+0.80
Palladium	+0.99
Platinum	+1.20
Gold	+1.50
Cobolt	+1.82

The standard conditions under which the electrode potential is measured are, 25°C and electrolytes of 1 molar solution.

Galvanic cells are classified into three groups; *composition cells, stress cells* and *concentration cells*.

Composition cells consist of two dissimilar metals in contact. The metal having the lower electrode potential becomes the anode and will corrode. Examples include galvanized steel (often used for manufacturing buckets) and tinplate. Galvanized steel consists of zinc coated mild steel sheet. Since zinc has a lower potential (−0.76 V) than the steel (Fe, −0.44 V), zinc becomes the anode that corrodes and so protects the mild steel. In the case of tinplate, which is a coating of tin (Sn, −0.14 V) on mild steel (Fe, −0.44 V), then iron becomes the anode which corrodes. The layer of tin provides a barrier to the corrosion, but if damaged, corrosion occurs. This is why damaged tin cans are removed from the shelves of shops, because corrosion of the underlying steel can contaminate the contents of the can.

Stress cells are formed where differences in stress within a material give rise to differences in electrode potential. For example regions within a component having different amounts of cold work during its forming process. The regions of strain hardening have a greater energy than annealed material and so become anodic and corrode. Thus stressed components are more likely to corrode than unstressed. More will be said about this when we look at the structure and cold working of materials.

An example where this may be a problem is with crimped metal joints, if the material is not given an appropriate heat-treatment after crimping, then high stress areas are created at the joints which will preferentially corrode.

Concentration cells arise due to differences of the ion concentrate in the electrolyte. The concentration cell accentuates corrosion, but it accentuates it where the concentration of the electrolyte is least. This relationship which occurs in practice can be proved theoretically.

In an oxidation type concentration cell, when the oxygen has access to a moist surface, a cathodic reaction can occur, such as:

$$2H_2O + O_2 + 4e^- \rightarrow 4(OH)^-$$

Thus regions where oxygen concentration is less become anodic and corrosion occurs (see Figure 3.15).

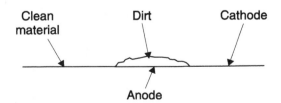

Figure 3.15 *Where the oxygen concentration is least becomes*

An example of this behaviour may occur on a car's bodywork, if a small amount of dirt is left on the bodywork where the surface may have been damaged, the area under the dirt will corrode. The oxygen in this area has been excluded i.e., it is least when compared to the clean surface adjacent to the dirty area. This is one of the reasons why we try and keep motor vehicles clean!

We have discussed the mechanisms of electro-chemical corrosion, so how as engineers' may we prevent it? There are in fact four major ways in which to prevent, or at least help reduce, the effects of corrosion. By protective coatings, cathodic protection, design and materials selection.

Protective coatings, form a barrier layer between the metal and the electrolyte. They must be non-porous and non-conductive. There are many types of coating suitable for this purpose. *Organic coatings* such as polymeric paints, *ceramic coatings* like enamels (these are brittle so care must be taken to avoid damage), metal coatings like tinplate and *chemically deposited coatings* such as the formation of a phosphate layer. Phosphating tends to be porous, but forms a keying layer for subsequent metal deposition or paint.

There is also the naturally occurring *protective oxide layer* that is formed by *passive metals*, which provides an excellent barrier against further corrosion. Aluminium, stainless steel, titanium and nickel are all examples of passive metals.

Cathodic protection is the mechanism whereby the anode is made to act as the cathode. It is achieved by use of either a sacrificial anode or an impressed d.c. voltage.

The sacrificial anode is a metal, that when connected in the form of a galvanic cell to the component being protected, forms the anode (Figure 3.16). Examples of galvanized steel where zinc is the sacrificial anode, magnesium or zinc plates attached to ships hulls, and buried pipes.

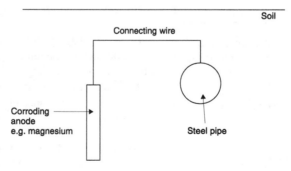

Figure 3.16 *Sacrificial anode*

For the impressed voltage (Figure 3.17), a d.c. source is connected to the metal to be protected and an auxiliary electrode, such that the electrons flow to the metal which then becomes the cathode. The auxiliary electrode has to be replaced from time to time.

There are numerous ways in which the designer can help prevent corrosion. If at all possible avoid the formation of galvanic cells, if dissimilar metals are to be brought into contact electrically insulate them. Try and ensure that ingress of moisture is avoided.

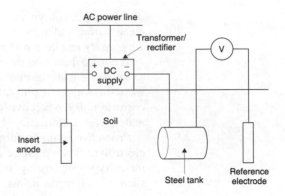

Figure 3.17 *Typical layout for impressed voltage corrosion protection*

Make the anodic area much larger than the cathodic area if possible, for example, use copper rivets to fasten steel sheet, not the other way round. When installing fluid systems, try and ensure that system components and plumbing is enclosed to avoid oxygen pick-up. Avoid areas of stagnant liquid. Finally, avoid pipework or components being internally stressed, remember these areas of high stress are more energetic and therefore become anodic and subsequently corrode.

Corrosion can also be minimized by materials selection and appropriate heat treatment, particularly in steels. More will be said later on the subject in the element concerned with materials selection.

Finally in our discussion on corrosion, a number of the more commonly encountered forms of corrosion are detailed below:

Pitting is a localized form of corrosion resulting in small holes that may completely penetrate some members. It is encountered in aluminium and its alloys, copper and its alloys, stainless steel and high-nickel alloys. Pitting is an electrochemical form of attack involving either galvanic or concentration cells, and sometimes both types of cell.

Intergranular corrosion, a form of galvanic attack, occurs when grain boundaries (see structure of materials) of a metal are selectively corroded. It is the result of composition differences between the grain boundary and the grains themselves.

High-temperature corrosion is accelerated by alternate heating and cooling because brittle protective scales expand and contract at different rates compared to the base metals supporting them. This causes flaking to occur exposing fresh metal to attack.

Stress-corrosion is likely to occur when surface tensile stresses act in combination with a corrosive medium. Failure is believed to start at the high energy grain boundaries, which are anodic when compared to the grains themselves.

Test your knowledge 3.6

Explain the terms: concentration cell, composition cell and stress cell, with respect to galvanic corrosion. As part of your explanation give examples where each type of cell may cause corrosion.

Test your knowledge 3.7

Zinc and copper form the electrodes of a galvanic cell.
(a) Which material is the anode and which is the cathode?
(b) Which material will corrode?
(c) Which way will the electrons flow around the external circuit?

Corrosion-fatigue is caused by the action of a corrosive medium combined with variable cyclic stresses. In this type of failure, a corrosive agent attacks the metal surface, where imperfections produce stress raisers that start a fatigue failure.

This study on the mechanisms and prevention of corrosion brings to an end our discussion on material properties. We will now look at the structure of materials at the macroscopic (molecular) level and see how this affects their properties.

Test your knowledge 3.8

Explain the difference between direct chemical attack and electrochemical attack, with respect to corrosion.

Test your knowledge 3.9

Under what conditions are the half-cell reactions of metals measured in order to produce the electrochemical series?

Activity 3.5

A material supplier, Beta Metals, has asked you to assist them in the production of some data sheets on corrosion protection that will be incorporated into the next edition of their catalogue. Identify four principal methods that can be used to protect against corrosion and, for each method, produce a summary that can be used to form the basis of the data sheets required by Beta Metals. Include relevant diagrams and sketches and make sure that your text is suitable for the non-technical reader.

Structure of materials

The chemical composition and the types of bond that hold atoms and molecules together are the major underlying characteristics that determine the properties of materials described in the previous section. We start therefore, by considering these bonding mechanisms. Then we study the structure of metals and alloys and the ways in which they behave under stress. Next we look at polymers and their various sub-divisions, finishing with a discussion on composite materials.

Bonding

As mentioned previously, the way in which atoms join together, or molecules join together, is called *bonding*. In order to fully understand the mechanisms of bonding you will need to be aware of one or two important facts about the atom and the relationship between the type of bond and the *periodic table*.

I am sure you are aware of the fact that the nucleus of the atom consists of an association of protons and neutrons and that the protons carry a positive charge. Surrounding the nucleus in a series of discreet energy bands, electrons (negative charge) orbit the nucleus. Electrons in the energy bands or shells closest to the nucleus are held tightly by electrostatic attraction. In the outermost shells they are held less tightly to the nucleus.

The *valence* of an atom is related to the ability of the atom to enter into chemical combination with other elements, this is often determined by the number of electrons in the outer most levels, where the binding energy is least. These valence shells are often known as s or p shells and the letters refer to the shell to which the electrons belong.

So, for example, magnesium which has 12 electrons, aluminium which has 13 electrons and germanium which has 32 electrons, can be represented as follows:

$$\text{Mg; } 1s^2 \ 2s^2 \ 2p^6 \ \underline{3s^2} \qquad\qquad\qquad \text{valence} = 2$$
$$\text{Al ; } 1s^2 \ 2s^2 \ 2p^6 \ \underline{3s^2 \ 3p^1} \qquad\qquad \text{valence} = 3$$
$$\text{Ge; } 1s^2 \ 2s^2 \ 2p^6 \ 3s^2 \ 3p^6 \ 3d^{10} \ \underline{4s^2 \ 4p^2} \qquad \text{valence} = 4$$

The numbers 1s, 2s, 2p, etc. relate to the shell level, the superscript numbers relate to the number of electrons in that shell. Remember the total number of s and p electrons in the outermost shell (those underlined) often accounts for the valence number. There is an exception to the above rule, the valence may also depend on the nature of the chemical reaction.

If an atom has a valence of zero, no electrons enter into chemical reactions and are all examples of inert elements.

Let us now turn our attention to the ways in which atoms combine or bond together. There are essentially three types of primary bond, *ionic*, *covalent*, and *metallic* as well as secondary bonds such as, *Van der Waals*.

When more than one type of atom is present in a material, one atom may donate its valence electrons to a different atom, filling the outer energy shell of the second atom. Both atoms now have full or empty outer energy levels but in the process, both have acquired an electrical charge and behave like ions. These oppositely charged ions are then attracted to one another and produce an *ionic bond*. The ionic bond is also sometimes referred to as the *electrovalent bond*.

The combination of a sodium atom with that of a chlorine atom illustrates the ionic bonding process very well, and is shown below in Figure 3.18.

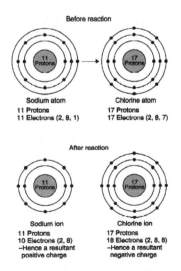

Figure 3.18 *Illustration of the ionic bonding process between a sodium and chlorine atom*

Note that in the *transfer* of the electron from the sodium atom to the chlorine atom, both the sodium and chlorine ions now have a noble gas configuration, where in the case of sodium the outer valence shell is empty while for chlorine it is full. These two ions in combination, are sitting in their lowest energy level and so readily combine. In this classic example of ionic bonding, the metal sodium has combined with the poisonous gas chlorine to form the sodium chloride molecule, common salt!

In *covalently bonded* materials electrons are *shared* among two or more atoms. This sharing between atoms is arranged in such a way that each atom has its outer *sp* shell filled, so that by forming the molecule each atom sits in its lowest energy level having a noble gas configuration. The covalent bonding of silicon and oxygen to form silica (SiO_2 silicon dioxide) is shown in Figure 3.19.

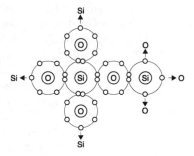

Figure 3.19 *Covalent bond formed between silicon and oxygen atoms*

The metallic elements that have low valence, give up their valence electrons readily to form a sea of electrons which surround the nucleus of the atoms. Thus in giving up their electrons the metallic elements form positive ions which are held together by mutual attraction of the surrounding electrons, producing the strong *metallic bond*. Figure 3.20 illustrates the metallic bond.

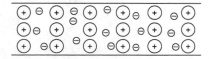

Figure 3.20 *Illustration of the metallic bond*

Van der Waals bonds join molecules or groups of atoms by weak electrostatic attraction. Many polymers, ceramics, water and other molecules tend to form electrical dipoles, that is, some portions of the molecules are positively charged while other portions are negatively charged. The electrostatic attraction between these oppositely charged regions weakly bond the two regions together (Figure 3.21).

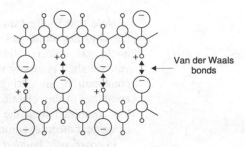

Figure 3.21 *Van der Waals bonds which join molecules or groups of atoms by weak electrostatic action*

Van der Waals bonds are *secondary bonds,* but the atoms within the molecules or groups of molecules are held together by strong covalent or ionic bonds. For example when water is boiled the secondary Van der Waals bonds that hold the molecules of water together are broken. Much higher temperatures are then required to break the covalent bonds that combine the oxygen and hydrogen atoms. The ductility of polyvinylchloride (PVC) is attributed to the weak Van der Waals bonds that hold the long chain molecules together. These are easily broken allowing these large molecules to slide over one another.

In many materials bonding between atoms is a mixture of two or more types. Iron for example, is formed from a combination of metallic and covalent bonds. Two or more metals may form a metallic compound, by a mixture of metallic and ionic bonds. Many ceramic and semiconducting compounds that are a combination of metallic and non-metallic elements, have mixtures of covalent and ionic bonds. The energy necessary to break a bond — the binding energy — is shown in Table 3.6.

Table 3.6 *Values of the binding energy for primary and secondary bonds*

Bond	Binding energy (kJ/mol)
Ionic	625–1550
Covalent	520–1250
Metallic	100–800
Van der Waals	<40

The *electronic structure* of an atom may be characterized by the energy levels to which each electron is assigned, in particular to the *valency* of each element. The periodic table of the elements is constructed based on this electronic structure.

The electronic structure plays an important role in determining the bonding between atoms, allowing us to assign general properties to each class of material. Thus metals have good ductility and electrical and thermal conductivity because of the metallic bond. Ceramics, semiconductors and many polymers are brittle and have poor conductivity because of their covalent and ionic bonds. While Van der Waals bonds are responsible for good ductility in certain polymers.

Structure of metals

We have already discussed how metals are formed from their atoms, by metallic bonding. The forces binding the atoms in a metal are non-directional, and since in a pure metal all atoms are of the same kind and therefore are the same size, they will arrange themselves in patterns to give the lowest potential energy. The patterns formed in this process are known as *crystals* or *grains* and the basic building block which replicates itself to form these crystals is known as the *unit lattice* or *unit cell*. Most metals of engineering importance form crystals by using one of three principal types of lattice structure. These are known as *body-centred cubic* (BCC), *face-centred cubic* (FCC) and *close packed hexagonal* (CPH).

The BCC lattice (Figure 3.22) packs the atoms together more loosely than in either the FCC or CPH systems. Since *plastic deformation* in metals is caused by atoms *slipping* over one another. Then the number of ways in which a structure allows this slip to occur, can help determine the degree of ductility of the metal.

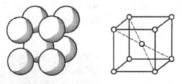

Figure 3.22 *Body-centred cubic unit cell*

The BCC structure has *few planes* over which slip occurs, when compared to either FCC or CPH structures. So metals with this type of structure tend to be less ductile, when compared with the others. This is shown to be true in practice since relatively brittle metals such as chromium, tungsten, molybdenum and vanadium all have a BCC structure.

In contrast metals having the FCC structure (Figure 3.23), where there are more opportunities for slip to occur, are relatively easily deformed and most metals with this structure display good ductility. Copper, gold, silver, aluminium and lead are all good examples of ductile metals with an FCC structure.

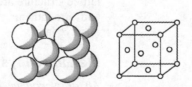

Figure 3.23 *Face-centred cubic unit cell*

Although the CPH structure (Figure 3.24) has a similar number of slip planes to FCC structures, due to the way in which the atoms are packed, relatively few of these planes actually promote slip (plastic deformation). So unexpectedly these metals display far less ductility than metal formed from the FCC lattice structure. Examples include beryllium, zinc and cadmium.

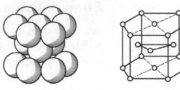

Figure 3.24 *Close packed hexagon unit cell*

Most metal products, involve, at some stage, melting the metal and casting it into moulds where it solidifies. The process of solidification favours the growth of many crystals (grains), each growing and interlocking as the molten metal cools and solidifies. The nucleation of these grains is thought to start from specks of dust or other minute impurities in the melt. The growth of an individual crystal (grain) is halted when its extremities come into contact with another grain, thus each grain is separated by a *grain boundary*. The process of grain growth as a metal cools from the melt is illustrated in Figure 3.25, in which the individual squares within each grid, represents the unit lattice or cell from which each grain is formed.

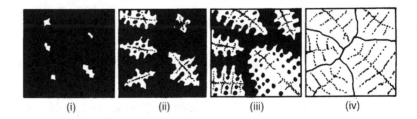

(i)   (ii)   (iii)   (iv)

Figure 3.25 *Grain growth as the metal cools from the melt*

You may remember that when the unit cell is formed it is made up in such a way as to ensure that the individual atoms within the cell are at their lowest energy level. Then any irregularity in the pattern of the lattice that forms the grain must raise the energy of the atoms. This occurs at grain boundaries where the regular grain structure is distorted, due to the orientation of the grains as they meet. The higher energy at grain boundaries and other imperfections in the lattice structure are important for our understanding of the hardening and strengthening mechanisms of metals and alloys, as you will observe later.

Metallic alloys

An alloy can be defined as the intimate mixture of two or more metallic elements in solid or liquid form. If we wish to modify the properties of pure metals, in order to meet engineering requirements, then we can alloy that metal with another and produce an alloy that maintains the best attributes of both. For example the alloy duralumin consists mainly of pure aluminium, mixed with about 4% of copper. Aluminium is very ductile but not particularly strong, by adding the copper the tensile strength of the alloy is significantly higher than the pure metal. Also with the appropriate heat treatment the ductility and toughness of the alloy can be made to approach that of pure aluminium. So, within limits, alloying gives us the ability to modify metals to suit our requirements.

Engineering alloys may be divided into two broad categories; ferrous alloys and non-ferrous alloys. As their name suggests the ferrous alloys are alloys of iron with small amounts of other elements added. The iron–carbon alloys or *steels* are very important engineering materials, which have been used successfully for many years. Their importance is emphasized by the fact that they are always considered as a separate set of alloys.

Non-ferrous alloys are broken down again into several important sub-divisions. These include; aluminium alloys, magnesium alloys, copper alloys, nickel alloys, zinc alloys and titanium alloys. Each sub-set taking its name from the principal alloying element.

Pure iron is an *allotropic* element, i.e. it can exist in more than one physical form or structure. At temperatures below 910°C the crystal structure of iron has a BCC lattice. Between temperatures of 910°C and 1400°C it exists as a FCC structure and above 1400°C it becomes BCC again.

These different physical states have common names. Iron in the BCC form is referred to as *ferrite*. Iron in its intermediate FCC form is referred to as *austinite*. Due to the difference in space between the atoms in the BCC lattice and FCC lattice they can accommodate varying numbers of carbon atoms. The FCC structure has the most room between atoms and so austinite in this form can accommodate more carbon up to around 2%, while, as ferrite, it can only accommodate up to 0.2%.

So as the alloy cools and the structure changes from FCC to BCC, the carbon comes out of the grains and forms a separate very hard and brittle compound known as *cementite*. So iron–carbon alloys having cooled from the solid may consist of pure ferrite grains or grains containing both ferrite and cementite, these mixed grains are often known as *pearlite*. As already mentioned the cementite is very hard and brittle, pearlite is less brittle, while ferrite is relatively soft and ductile. The properties of iron alloys vary smoothly and continuously with changes in carbon content, because as the carbon content changes so does the structure of the alloy.

Wrought iron, used for making ornamental gates and other items that require a very malleable metal, is an alloy that contains only minute amounts of carbon, hence consists almost exclusively of the very ductile and soft ferrite grains, so is ideal for forging into shape.

Mild steel has around 0.1% carbon and is malleable enough to be formed by pressing or drawing operations (see materials processing), being ductile its used for lightly stressed parts. Commercial tool steels contain from 0.7% to about 1.3% carbon and as their name suggests are used for hand tools like chisels, files taps and dies, etc. Cast iron which is a particularly brittle material contains from about 2% to 4.5% carbon, it has good machining properties and is extremely hard so is often used for items like surface tables, heavy machine castings, motor vehicle cylinder blocks, etc.

If we add small amounts of other elements, apart from carbon, to iron, we can produce a range of steels with all sorts of useful properties. For example, small amounts of chromium, magnesium, nickel and titanium added to iron and carbon can produce a range of stainless and heat resisting steels. These may be used for kitchenware, clinical and hospital utensils, and special non-toxic pressure vessels used for example, in the brewing industry.

Example 3.5

Explain why cast-iron is an ideal material for the manufacture of engine cylinder blocks.

Engine cylinder blocks need a smooth warp free precision base on which to place the cylinder head containing the valve gear, as well as being hard enough to withstand the continuous rubbing caused by the piston rings. Cast-iron is a dense material that is virtually warp free, even at elevated temperatures. It is extremely hard and so has good resistance to abrasion. It has excellent machining properties and so is ideal for producing close tolerance precision surfaces, and finally, it is easily cast into complex

---

Non-ferrous alloys as their name suggests involve metals other than iron as the primary alloying element. The names of some of these alloys have already been mentioned, one special set of these alloys are known as *light alloys*. These are alloys of beryllium, aluminium, magnesium and titanium where the densities range from 1700 kg/m^3 to 4500 kg/m^3 which compare with densities of 7800 kg/m^3 to 8900 kg/m^3 for iron and copper.

The property of lightness has led to the association of light metals with transportation and more especially with the aerospace industry which has provided a great stimulus to the development of these alloys during the last 45 years. Their good structural efficiency in terms of strength and stiffness, has been dominant in their choice by designers.

Materials testing

Destructive tests are used to determine certain mechanical properties of materials, for instance the common *tensile test* may be used to check the strength and ductility of a material. Test specimens may be obtained by taking samples from a batch on a production line or using laboratory specimens that have been especially prepared to check the properties of a new material. Other material properties that can be determined from destructive tests include, hardness (perhaps checking the effectiveness of a heat treatment), creep, fatigue and toughness.

Non-destructive testing can be used to determine the integrity of components in service, or to confirm that materials have been processed correctly. Ultrasonics is one method of non-destructive testing which is particularly suitable for determining whether or not the layers of carbon fibre cloth have been fully impregnated by the resin matrix, when manufacturing a carbon fibre reinforced plastic component.

We will now consider some of the more common destructive and non-destructive tests. Later on you will be asked to carry out some of these tests, analysing and comparing the data obtained from them.

Tensile testing

The tensile test is one of the most important materials property tests. It enables us to determine tensile strength, elastic modulus, yield strength, maximum percentage elongation (and thus determine Poisson's ratio) and, it tells us a lot about the ductile behaviour of materials. It is particularly suited for testing metals and many polymers. In order to ensure that the tests are accurate specimen sizes should be of set dimensions which conform to BSEN 10002 or other related standards produced by the British Standards Institute. If tests are to be carried out using non-standard equipment, then the appropriate standard should always be consulted for guidance.

Standard tensile specimens may be rectangular or circular in cross-section, as shown in Figure 3.26. The gauge length should be $5 \times$ (diameter) in the case of round specimens and $5.64 \times$ (cross-sectional area)$^{1/2}$ for rectangular ones.

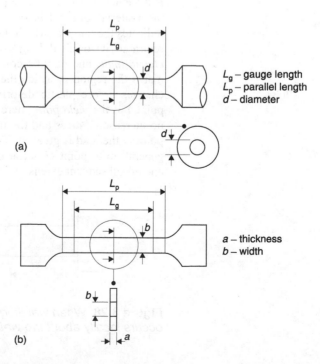

Figure 3.26 *Typical standard tensile test specimens which can have round or rectangular cross-sections that conform to BSEN 10002*

Tensile testing machines range from small hand-operated bench models known as *tensometers,* to large floor-standing machines that are capable of exerting forces of several hundred kilonewtons on the specimen. As the specimen is loaded the corresponding extensions, are recorded either by taking readings, or graphically by indenting specially prepared graph paper with a pointer which corresponds with the load and extension imposed at the time on the specimen. A typical load–extension graph for a *mild steel* is shown in Figure 3.27.

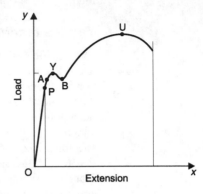

Figure 3.27 *Typical load–extension graph for mild steel*

The graph of Figure 3.27 is typical for a ductile material like mild steel. The key phases resulting from the test are clearly marked on the graph. The point *P* is the *limit of proportionality.* Between *O* and *P* the extension *x* is directly proportional to the applied force *F* and the material obeys Hooke's law, which was defined earlier. *OA* indicates the range in which the specimen extends elastically and will return to its original length the *elastic stage*. The point *A* is the *elastic limit* and may be coincident with *P* for some materials. *OY* marks the region in which the material undergoes internal structural changes before plastic deformation and permanent set occur. The point *Y* is the *yield point*, here the extension suddenly increases with no increase in stress and the material enters the *plastic stage*. At the point *U* the load is greatest. The extension of the test piece has been general up to point *U* but at point *U waisting* occurs (Figure 3.28), and all subsequent extension is local.

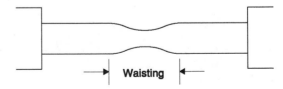

Figure 3.28 *When waisting occurs all subsequent extension occurs locally about the waist*

Since the area at the waist is considerably reduced then from stress = force/area, stress will increase, resulting in a falling-off of the load, and *fracture* occurs beyond point *U* in Figure 3.27.

If the length and diameter of the specimen is taken prior to and after the test, then the percentage elongation and reduction in cross-sectional area (csa) can be determined from the following relationships.

$$\text{percentage elongation} = \frac{\text{increase in gauge length}}{\text{gauge length}} \times 100\%$$

and

$$\text{percentage reduction in area} = \frac{\text{original csa - final csa}}{\text{original csa}} \times 100\%$$

These values can now be used to calculate Poisson's ratio for the material, remembering our definition! Also from the graph or from the readings taken we are able to determine the elastic modulus (stiffness) of the material.

Figure 3.29 provides some further examples of typical load–extension curves for a variety of materials.

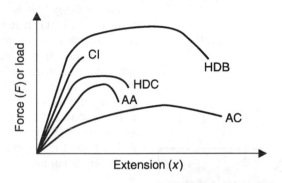

Figure 3.29 *Typical load–extension graphs where HDB = hard drawn brass, CI = cast iron, HDC = hard drawn copper, AA = aluminium allow, AC = annealed copper*

Figure 3.29 shows that annealed copper is a very ductile material, whilst hard drawn copper is stronger but less ductile. Hard drawn 70/30 brass (copper–zinc alloy) is both strong and ductile. Cast-iron can clearly be seen as brittle and it is for this reason that cast-iron is seldom used under tensile load. Aluminium alloy is seen to be fairly strong yet ductile. Although not indicated here aluminium is one of the light alloys (magnesium and titanium being examples of others) that is strong yet light, i.e. it has a very good structural efficiency. Structural efficiency is defined as:

$$\text{Structural efficiency} = \frac{\text{yield stress}}{\text{density}}$$

or using symbols

$$\eta = \frac{\sigma_y}{\rho}$$

Some other properties of importance to the engineer are given below, all of which can be determined from the results of the tensile test.

Working stress is the stress imposed on the material as a result of the worst possible load experienced in service and, must be within the elastic limit.

Proof stress is the tensile stress which when applied for a short period of time (often in the region 0 to 15 seconds), produces a permanent set of a specified amount, usually 0.1%. Do not mix this up with the yield stress which has already been defined.

Ultimate tensile stress is given by the relationship:

$$\frac{\text{maximum load}}{\text{original cross-sectional area}}$$

Note that the point U on the load–extension graph, shows the maximum load. This must be divided by the original csa not that directly under the point U.

Factor of safety   It is common design practice to use the tensile strength of a metal that is easily obtained from tensile tests, and a *factor of safety* to ensure that uncertainty is taken into account, so that the stress stays within safe limits. In determining a factor of safety consideration will need to be given to such things as; the uniformity of the chosen material; the type of loading-static or dynamic; the likely effect of failure; the effect of wear or corrosion of the material and the environment in which the material is to operate. Factors of safety can vary from about 1.5 to 3 for static loads, to about 15 for impact loads or 20 where fluctuating loads may cause fatigue failure.

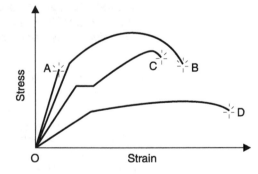

Figure 3.30 *Stress–strain curves for four common materials*

Hardness testing

The hardness test measures the resistance of a material to indentation. A load is gradually applied to an indenter, which is positioned at right angles to the test material. The test causes the material to plastically deform. The properties that affect plastic deformation are identical to those that affect hardness and so the test is a good measure of the hardness of the material. There is also a link between the hardness of a material and its tensile strength, since

---

Test your knowledge 3.10

Figure 3.30 shows stress–strain curves for four common materials.
(a) Which material has the highest yield stress?
(b) Which material is the stiffest?
(c) Which material is the strongest?
(d) Which material has the greatest ductility?

Test your knowledge 3.11

Define the following terms, with respect to tensile testing:
(a) Elastic deformation
(b) Plastic deformation
(c) Ultimate tensile strength
(d) Proof stress
(e) Percentage elongation
(f) Poisson's ratio.

hardness is measured as the indenter force divided by the projected area of the indenter, i.e.

$$\text{Hardness} = \frac{\text{indenter force}}{\text{surface area of indenter}}$$

$$= \frac{F}{A} \text{ (same units as stress)}$$

and for some materials, particularly steels, the tensile strength measured in N/mm^2 is approximately three times the hardness number.

The three most common types of hardness test are the *Brinell, Vickers* and *Rockwell* tests.

In the *Brinell* hardness test a steel or tungsten carbide ball is used as the indenter (Figure 3.31). This is forced into the material under load for about 15 seconds. The average diameter d of the indentation made is then measured in millimetres using a microscope. A ball diameter of 10mm is often used. The applied loads vary according to the nature of the material under test, but do not normally exceed 3000kg.

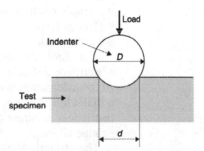

Figure 3.31 *The Brinell hardness test uses a hardened steel ball indenter*

The Brinell hardness number *HB* is then defined as:

$$HB = \frac{\text{load } (F)}{\text{surface area of indentation}} = \frac{2F}{\pi D \left[D - \left(D^2 - d^2 \right)^{0.5} \right]}$$

If the indentation force F is measured in Newtons instead of kg force, then the above formula must be multiplied by a factor of 0.102.

In the Vickers hardness test the indentation produced is very small and needs to be measured by a microscope. Here the indenter used is a diamond square based pyramid (Figure 3.32). Test loads will again vary (normally 5 to 100kg) according to the nature of the material under test. The Vickers hardness number (VHN) for a load F kg is given by:

$$VHN = \frac{1.854}{d^2} \text{ where } d = \frac{d_1 + d_2}{2}$$

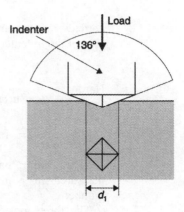

Figure 3.32 *The Vickers hardness test uses a diamond square-based pyramid indenter*

In practice there is no need to calculate the hardness number with either the Brinell or Vickers tests. Tables are provided which can be related to the measurements taken, at the time of the test, which also take into account the load used and the size of the indenter. There are no simple relationships between the hardness scales obtained for the different tests, because different methods are used to obtain a hardness value.

The *Rockwell* test uses either a hardened steel ball for softer material or a diamond cone indenter for harder materials. The depth of the indentation is converted to a hardness reading that is shown on a dial while the material being tested is still under load. This type of test is less accurate than the Vickers test but is still useful for rapid routine checks.

The Brinell, Vickers and Rockwell tests can all be used, quite successfully, for determining the hardness of plastics.

Bending tests

Bending tests are used to estimate the ductility of a material. The material is examined for cracks after it has been bent through some specified angle. There are several forms of bending test, all of which need to comply with the standards given in BS 1639. The tests may be carried out with or without the use of a former. Figure 3.33(a) illustrates a guided bending test where the radius of bend is controlled using a former, Figure 3.33(b) illustrates a free bend test where the material has been bent through 180°.

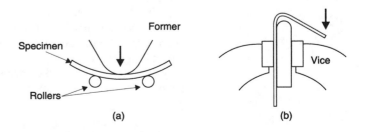

Figure 3.33 *(a) Guided bend test, (b) Free bend test*

Toughness testing

Impact tests are used to measure the energy required to fracture a standard notched bar using a heavy mass which is released at height and allowed to swing freely, in an arc to strike the test piece. The principle of operation of a typical standard impact test machine is illustrated in Figure 3.34.

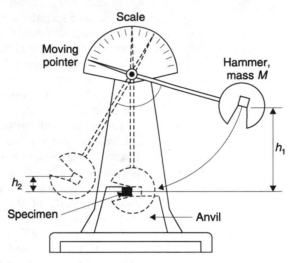

Figure 3.34 *Principle of operation of a typical standard impact test machine*

The potential energy of the mass can be determined mathematically prior to release and compared with the energy of the mass after impact, the difference in these two energies is the amount of energy absorbed by the test specimen and is a direct measure of the toughness of the material.

There are two common types of impact test, the Izod and the Charpy, they differ only in the way in which the test specimen is supported and the test specimen geometry. In the Izod test the specimen used is a circular or square-sectioned cantilever, with a V-notch mounted vertically in the jaws of the anvil as shown in Figure 3.35(a).

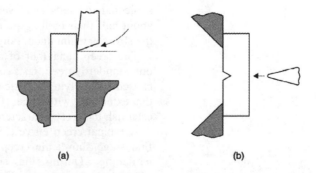

Figure 3.35 *(a) Cantilever impact test specimen, (b) Beam impact test specimen*

In the Charpy test a simply supported square sectioned beam is used. Again containing a specially prepared V-notch, but this time the notch is mounted on the far side from the point of impact (Figure 3.35b), so that the specimen fractures at the notch.

Impact testing can be used for both metals and polymers, the fracture surface will vary according to whether the material is brittle or ductile. Brittle metals show a clean break, with little change in cross sectional area and little or no plastic deformation. The fracture surface has a granular structure.

With ductile metals the fracture surface is rough and fibrous. The more ductile a material the more energy is required to cause fracture. The energy of impact is absorbed in causing plastic deformation within the material. Ductile test pieces tend to bend rather than completely break, plastic flow can be identified by noting the corresponding reduction in cross sectional area.

The results with brittle and ductile polymers are similar to those for metals. With brittle polymers there is a clean break and the fracture surfaces are smooth and glassy in appearance. With ductile polymers there is a significant reduction in cross-sectional area, with little or no fracture having taken place. With polymers that contain a colour pigment, at the point of impact, plastic straining of the material can be seen by a reduction in colour giving a translucent (semi-transparent) effect.

In both the Izod and Charpy tests the energy absorbed at impact is measured to given an indication of toughness of the material under test. For metals this is simply expressed in Joules (J). With polymers the width of the notch or the cross-sectional area of the test piece are also considered, so the toughness of these materials may be expressed in units of (J/m) or (J/m^2), dependent on the type of test.

Other destructive tests

Apart from the forms of destructive testing already mentioned materials can be subjected to *shear* tests. Where the specimen in the form of a thin-walled cylinder is subject to a twisting couple. This twisting moment simulates a shearing action within the walls of the test piece, from which we can determine information on the shear strength (symbol τ) and the shear modulus or modulus of rigidity (symbol G) of the test specimen.

Creep has already been defined as the slow plastic deformation of a material accelerated with increase in temperatures. Metals when subjected to loads at elevated temperatures (normally greater than about half their melting point) may continue to deform slowly while the stress is maintained. Polymers can creep at room temperatures.

The creep behaviour of materials can be determined by carrying out standard creep tests on material specimens. Here the creep caused by applying a static tensile load, is determined by measuring the extension with time. The results are plotted and analyzed to establish the creep characteristics of the material.

A typical creep curve (Figure 3.36) consists of three stages. The first stage shows the creep rate being gradually reduced by strain hardening. During the second stage the creep rate remains approximately constant. Finally in the third stage there is a rapid increase in the strain rate, due to necking, until ultimately fracture occurs.

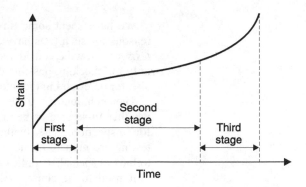

Figure 3.36 *Typical creep curve with specimen subject to constant stress at constant temperature*

In order to establish the fatigue strength of a material, specimens are subjected to some form of dynamic cyclic stressing in a fatigue testing machine. The type of test will depend on the form of the material, i.e. whether in the form of sheet, strip, bars, shafts, etc. Specimens can be pulled and pushed to simulate continuous tensile and compressive loading, or they may be rotated (twisted) backwards and forwards again to set up cyclic loads.

Interpretation of the results of fatigue tests can be complicated, but are necessary because materials may fail at much lower stress levels than when subject to static loads only. The *endurance limit* for steels and the *fatigue strength* for other materials such as light alloys can be determined. The designer can then ensure the life of chosen materials subject to fatigue loads, provided these loads are not exceeded.

Steels subject to cyclic loading will not fail by fatigue, providing this *endurance limit* is not exceeded (Figure 3.37). Other metals such as the light alloys do not have an endurance limit, so no matter how low the cyclic loading is eventually, given enough cycles, the material will fail by fatigue. By knowing the magnitude (size) of the maximum cyclic loads imposed on a material then a fatigue life can be estimated either directly as a number of cycles or, by using statistics, a time in hours or years is established. This technique, which enables a material fatigue life to be estimated, is particularly useful for aircraft designers where the problems created by fatigue are of major importance.

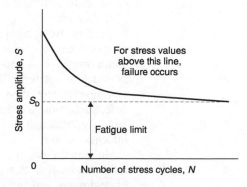

Figure 3.37 *Typical S/N curve for a steel showing that there is a fatigue endurance limiting stress*

We have spent some time discussing destructive testing, and the reasons for such tests have already been mentioned. *Non-destructive tests* will now be considered, remembering that these are useful, not only for checking possible defects produced during manufacture, but also for determining the integrity of materials being used in-service.

The technique used will depend on the nature of the material, whether in-service or being manufactured and the degree of access for inspection. The methods now available for non-destructive testing are numerous and whole fields of research have been set-up to look at specialist applications for any one individual method.

It used to be considered that there were five major methods – radiographic, ultrasonic, magnetic, electrical and penetrant. All of these methods having sub-divisions. Now, in addition, a range of new techniques have been developed including; acoustic emission methods, thermography, and holography.

Visual inspections, although not considered separately, are still of great importance, whether using the naked eye or optical aids such as boroscopes. We will not be able to cover all these methods here, so I will concentrate on one or two of the more important conventional methods, listing others and indicating their uses.

Visual inspection

Visual inspection can provide useful information on the surface condition of a material. Boroscopes (low powered microscopes) have been designed with the aid of fibre optics to help see into confined spaces, that are inaccessible to the naked eye. Aircraft engine combustion chambers that need regular inspection for defects, provide us with an example of where use of the boroscope is particularly helpful.

Care must be taken to ensure that no material defects have been overlooked. So this method is limited to detecting defects that are of sufficient size as to be readily seen. In aluminium alloys and low strength steels fatigue cracks do not generally become critical until they exceed about 2 to 3 cm in length, they are therefore capable of being detected visually before they cause failure of the structure. So for these materials visual methods of non-destructive testing are appropriate. This would not be the case when inspecting high strength steels or other more brittle materials, where the critical crack length can be less than 0.5 mm, here we would need to consider a more sensitive technique such as ultrasonics or radiography.

Holography is a technique that fits broadly under visual inspection. Holographic techniques can be used for the comparison of specimens, or the measurement of small amounts of deformation under stress, or a study of the surface during vibration. As the presence of a defect is likely to cause changes in the deformation or vibration pattern then by definition, holography can be used for defect detection. Normal photographic film is sensitive only to light intensity, although the image being photographed has much more information in it.

Holography not only uses the intensity of light, but also its phase information. It is this phase information that provides us with the three-dimensional image, which can be stored after processing and reproduced as a virtual image when illuminated in a particular way.

Radiography

When a beam of X-rays or more energetic gamma-rays are passed through a material onto a photographic plate, an image is obtained. The photographic plate is exposed to the rays for a period of time, which depends on the intensity of the rays, the thickness of the specimen and the characteristics of the film. The film is then processed and placed on an illuminated screen, so that the image can be examined and interpreted. Figure 3.38 shows a typical arrangement for flaw detection using X-ray radiography.

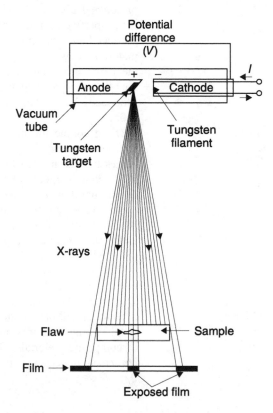

Figure 3.38 *Typical arrangement for flaw detection using X-ray radiography*

X-ray radiography requires a power source so the equipment is rather complex and cumbersome. Radiographical techniques are very useful for examining castings for porosity, blowholes, shrinkage cavities, cracks and other flaws. Gamma-ray radiography does not need an external power source because the rays are produced directly from a radioactive material such as Caesium-137, Cobalt-60 or Iridium-92, dependent on the intensity of the gamma-rays required. The portable equipment needed is therefore much less cumbersome than that required for X-ray radiography.

Radiography can be used for detecting flaws or defects in metals, and non-metals including composites. All forms of radiography are dangerous and particular care must be exercised when using the equipment, so that nobody is inadvertently exposed to radiation.

Ultrasonic testing

Mechanical vibrations can be made to travel through solids, liquids and gases. The actual particles of matter vibrate. If the frequency of vibration is within the range 16–20 kHz (16 to 20,000 cycles per second) the sound is audible to humans. Above 20 kHz the sound waves are referred to as ultrasound or ultrasonics. Typical frequencies used for ultrasonic testing are in the range 500 kHz to 20 MHz.

There are a number of ways in which ultrasonic waves can be produced but for non-destructive testing, most equipment uses the *piezoelectric effect*, for the test probes. A piezoelectric material has the property that if deformed by external pressure electric charges are produced and conversely, that if an electrical source is applied to the material it will change shape. By using alternating current a mechanical oscillation is produced in the piezoelectric plate. By coupling the piezoelectric probe to the specimen, sound waves are produced that match the frequency of vibration produced at the probe.

If we couple the probe to the specimen being tested a pulse of ultrasonic energy is generated within the material. This pulse of waves travels through the material with some spreading and will be reflected or scattered at any surface or internal discontinuity, such as an internal flaw in the specimen (Figure 3.39).

This reflected or scattered energy can be detected by a suitably placed second piezoelectric disc (receiver) on the material surface and will generate a pulse of electrical energy as a result of the received vibrations. The time interval between the transmitted and reflected pulses is a measure of the distance of the discontinuity from the surface and the size of the return pulse can be a measure of the size of the flaw. This is the simple principle of the ultrasonic flaw detector and ultrasonic thickness gauge.

Ultrasonic examination is particularly suited for the examination of composite materials. Disbonds and delamination (see structure of composite materials) can be identified, after manufacture or repair.

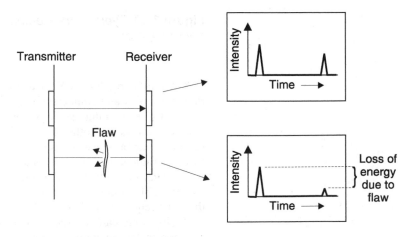

Figure 3.39 *Typical arrangement for ultrasonic flaw detection*

Activity 3.6

Beta Materials has asked you to carry out non-destructive tests on the following materials:

(a) a low carbon steel
(b) a carbon reinforced plastic composite material
(c) an aluminium alloy aircraft skin
(d) a large casting produced from cast iron.

For each material, list two techniques that you consider would be suitable and give reasons for your choice. Present your recommendations in the form of a word processed report.

Activity 3.7

Investigate the acoustic emission and thermographic methods of non-destructive testing. Write short notes describing each method and include:

(a) the physical principles used
(b) their advantages and disadvantages compared with other methods
(c) their industrial uses.

Activity 3.8

Use your library to investigate the procedure for the dye-penetrant method when used to detect possible surface defects on a steel component. Use the information obtained to write a brief explanation of the method and also give an account of how the results of the test may be interpreted.

Materials, characteristics and processing methods

In this element we are going to consider the processing of materials and the implications that these processing methods have on their property values. We will concentrate on a selection of metals, polymers and composite materials. Determining how, by modifying their grain structure, composition or by heat treatment, we are able to alter their properties.

Metals

Crystal structure, cold working and heat treatment

We have already learnt about how the atoms of metals bond together and form regular *lattice structures*. When a metal cools from the melt these lattice structures repeatedly join together three-dimensionally and grow *dendritically* to form *crystals* or *grains* (Figure 3.40). It is the size and nature of these grains that very much determines the properties of the parent material. Alloying metals also modifies their behaviour, during the manufacture of such alloys the metallurgist will try to maximize the good qualities that the individual elements bring to the alloy being produced.

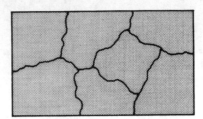

Figure 3.40 *Grains formed on solidification of the metal*

The strength of metals and alloys can be improved by modifying their grain structures. For pure metals an increase in yield stress can be achieved by reduction in grain size or by cold working that not only increases the yield stress but also increases hardness.

A regular structure within the grains of a metal helps to maintain that metal in its lowest energy state (see bonding of materials). When this regular pattern is disturbed by either defects or an increase in grain boundary area, the internal energy of the metal is increased. Thus the energy needed to fracture the material is also increased. This increase in fracture strength is measured by a resulting increase in yield stress.

Therefore an increase in yield stress is achieved by reducing the grain size and so increasing the high energy grain boundary area (Figure 3.41).

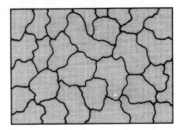

Figure 3.41 *Increase in grain boundary area as a result of reduction in grain*

If we cold work a metal say by cold-rolling, we increase the dislocation density within the grains. That is we raise the energy of the metal by increasing the amount of line defects (dislocations) present. Further plastic deformation (resulting from lines of atoms slipping over one another) is impeded due to the increase in energy required to overcome the increased density of dislocations (line defects) that are already present due to the cold work. Figure 3.42 illustrates this idea.

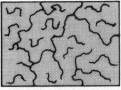 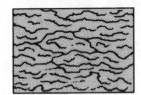

Dislocations (line defects)
prior to cold work

Figure 3.42 *The increase in dislocation density as a result of cold-working the material*

As we have already said an increase in cold work produces a subsequent increase in the number of dislocations which results in them becoming entangled with one another. It is these *dislocation tangles* that impedes the progress of further line defects and so plastic deformation only occurs at much higher stress levels. At the same time, for the same reasons, cold-working also produces an increase in hardness.

When a cold-worked metal is heated to about a third of its melting temperature (measured on the Kelvin scale), then there is a marked reduction in tensile strength and hardness. What actually happens is that at this temperature recrystallization takes place (Figure 3.43). Further increase in temperature enables the crystals to grow until the original distorted crystals, resulting from the cold work, are all replaced. The term *annealing* is used for the heat treatment process whereby the material is heated to above the recrystallization temperature and more ductile properties replace those produced by the cold work. The greater the amount of cold work the smaller the grain size produced after heat treatment, with a subsequent increase in yield stress when compared to a metal which has not been cold worked.

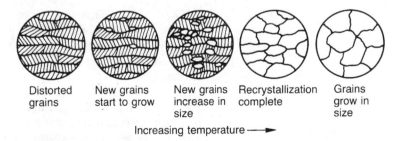

Distorted grains   New grains start to grow   New grains increase in size   Recrystallization complete   Grains grow in size

Increasing temperature ⟶

Figure 3.43 *Illustration of the recrystallization process with increasing temperature*

If copper is cold rolled then the yield stress and hardness will increase. At about 60% cold work the copper becomes so brittle that any further cold work results in fracture. So if further reduction in cross-section is required then the copper needs to be *process annealed* the original ductility and malleability is restored and the cold working process may continue.

Cold working involves plastically deforming the material at temperatures below the recrystallization temperature. Hot working as you might expect involves deforming the material at temperatures above the recrystallization temperature. So that as the grains deform they immediately recrystallize thus no hardening takes place and processing can continue without difficulty. Thus processing often involves hot working initially to produce the maximum amount of plastic deformation followed by cold rolling to improve the surface finish and increase the surface hardness as required. For example aluminium baking foil may be produced from blocks of aluminium in billet form, which are first hot rolled then cold rolled to produce thin even sheets.

Increase in yield stress for metal *alloys* is again achieved by modifying the grain structure of the host metal, but in a different way to that described above. Apart from line defects within the grain there may also be *point defects*. The various types of point defect are illustrated in Figure 3.44.

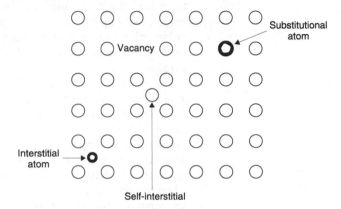

Figure 3.44 *Various types of point defect. A vacancy self interstitial is known as a Frenkel defect. The simple vacancy on its own is known as a Schottky defect.*

These point defects may be caused by impurities in the melt or may be deliberately introduced by *alloying*. In either case the net result is to increase the energy within the grains or at the grain boundaries, resulting once again in increases in yield stress and hardness.

The above process is particularly useful in improving the properties of aluminium alloys, and other light alloys. The heat treatment process involved is known as *precipitation hardening*. The alloy is heated to above its recrystallization temperature, then quenched. The result is a distorted high energy lattice structure, where the alloying elements initially lodge within the grains. After a time the fine precipitate diffuses toward the grain boundaries where it stays. Once again the grain boundary energy is increased making slip very difficult, hence an increase in yield stress (a measure of strength) and an increase in hardness.

A classic example of this process is found in the Al–Cu alloy *duralumin* (named after the town Duran in Germany where it was first produced). In the annealed state this alloy has a tensile strength in the region of 180–190 MPa and a hardness around 40–50 HB. After precipitation hardening these values may climb to around 420 MPa and 100 HB, respectively. This alloy is often used in aircraft skin construction, where it is clad with pure aluminium to improve its corrosion resistance properties while at the same time increasing its strength and hardness over and above that of pure aluminium.

Steels may be made very hard by heating them to above their recrystallization temperature and then suddenly reducing their temperature by quenching in water or oil. The sudden reduction in temperature does not allow sufficient time for the carbon atom to diffuse to the grain boundaries and form cementite. Instead the excess carbon atoms are trapped within the lattice structure and form a separate very hard phase known as *martensite*. This very hard and brittle substance has the effect of increasing the hardness and brittleness of the steel. The above process is known as *quenching*. Quenched steels may be up to four times as hard as their annealed counterpart, dependent upon the original amount of carbon added to the alloy. For example, for a steel produced from 1% carbon, which in its annealed state has a hardness value of 200 HV, once quenched this figure maybe as high as 800 HV.

After hardening a steel by quenching some of its ductility and resilience may be restored by *tempering*. Where the steel is reheated so that the carbon atoms can diffuse out and reduced the distortion within the lattice thus reducing the energy stored and so returning some ductility back to the steel. The higher the tempering temperature the more ductile the steel.

Processing metals

Metals may be processed or fabricated by two major methods, either *shaped* or *joined*, to form a component or structure. Metals may be shaped into something approaching the final form by one of several operations including, casting, rolling, drawing, forging, extruding, cutting, grinding and sintering. Or fabricated by joining with adhesives, fasteners, soldering, welding or brazing. Time does not permit all of the processes to be covered comprehensively in this text. We will, however, concentrate on a few processing techniques leaving you to research further processes as an exercise .

Casting of metals requires the metal in liquid form to be poured into a mould and then allowed to solidify before breaking the mould open to reveal the cast product. In order to ensure that the liquid metal reaches all parts of the mould, we either choose alloying elements that provide a low viscosity alloy (that is an alloy which easily flows) or, we use some external pressure to force the molten metal into the mould, as in pressure die casting.

The grain structure of the metal, within the mould, is dependent upon the rate of cooling. If the metal is rapidly cooled only small crystals have time to form. As the cooling rate is reduced so the crystal size grows. Metals that cool within a mould may cool quickly near the surface of the mould and due to heat flow may form long crystals which grow towards the centre of the mould (columnar crystals). At the centre of the mould due to thermal convection the moulten metal is continually on the move, this results in an even heat distribution, which results in the production of equal sized round crystals (equiaxed) being formed. The resultant structure of cast product moulded in an ingot is illustrated in Figure 3.45.

The method used for casting metals will depend on the size, number and complexity of the final product. Sand casting is suitable for small batch production of large components. If large numbers of castings are required a far superior product, with better dimensional

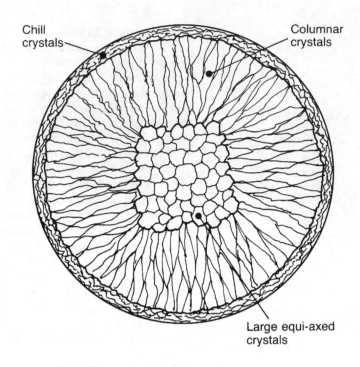

Chill crystals

Columnar crystals

Large equi-axed crystals

Figure 3.45 *Resultant crystal structure of a casting moulded in an ingot*

accuracy, is produced by using a metal mould. In permanent mould casting known as *gravity die casting*, the molten alloy is allowed to run into the mould under gravity, while in *pressure* die casting the charge is forced into the mould under considerable pressure.

The use of die casting is confined mainly to aluminium or zinc based alloys. Metal dies are expensive to produce, so that die casting in all its forms is only economically viable for large scale production. Pressure die casting produces components with good dimensional accuracy, uniform grain structure and good surface finish. So the metallic components produced by this method require little or no further processing. Sand casting does have the advantage of producing intricate shapes, because of the possibility of using destructible cores.

Since most metals become considerably softer and more malleable as temperature rises less energy is needed to produce a given amount of deformation, this is why *hot-working* is extensively used to shape metals. Drop-forging involves the use of a shaped die, one half being attached to the hammer and other to the anvil (Figure 3.46). When producing complex shapes by this method a series of dies may be used.

The hammer may be operated mechanically, pneumatically or hydraulically, dependent on the nature of the task.

The *extrusion* process is used for shaping a variety of ferrous and non-ferrous metals and alloys. The metal billet is heated to the required temperature, the ram is then driven hydraulically with sufficient pressure to force the metal through a hardened steel die. The solid metal section exudes from the die in a similar manner to that of toothpaste being squeezed from a tube. Figure 3.47 illustrates this process.

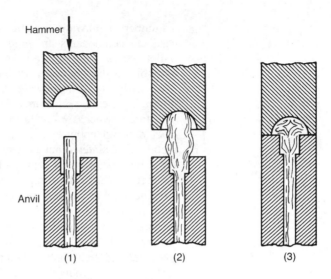

Figure 3.46 *Typical forging process showing resultant structure*

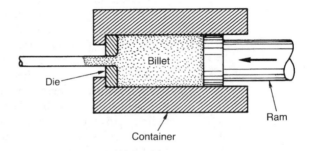

Figure 3.47 *An extrusion process*

The extrusion process is able to produce a wide variety of sections, including round and hexagonal rod, curtain rails, tubes, bearing sections and ordinary wire.

Cold-pressing and deep-drawing are closely related to each other and it is difficult to differentiate between them. The operations range from making a suitable pressing in one stage to cupping followed by

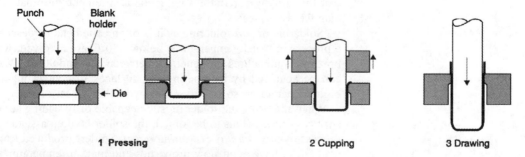

Figure 3.48 *A deep drawing process*

a number of drawing operations as shown in Figure 3.48. In each case the components are produced from sheet metal. Car bodies, bullet cases and general metal containers are examples of components that are easily produced using this process.

Sintering from a powder has become an important way of producing metallic structures. The metals to be sintered, in the form of a fine powder, are mixed together and then placed in a hardened steel die and compressed. The pressures used depend upon the metals to be sintered but are usually between 70 and 700 MN/m^2. At these high pressures a degree of cold welding takes place between the metals. The brittle compressed mass is then heated in a furnace to a temperature at which sintering (grain growth across the cold welds) takes place.

Tungsten is compacted and sintered in this way and the resulting sintered rod may be drawn into a fine wire to produce tungsten filaments for light bulbs. Tungsten carbide products for machine tools are also produced in this way. Cobalt being used with the tungsten carbide to produce a tough shock resistant bonding agent between the particles.

Sintering may also be used to produce bronze bearings. Here copper, tin and graphite are used in the sintering process to produce self-lubricating bearings.

Let us now turn our attention to just one metal cutting process, that of *machining*. This is essentially a cold working process in which the cutting edge of the tool forms chips or shavings of the material being machined. Very ductile alloys do not machine well, because local fracture does not occur ahead of the cutting tool edge. Thus brittle materials are considered to have good machining properties. Ideal materials have a suitable concentration of small isolated particles in their microstructure. These particles have the effect of setting up local stress raisers, as the cutting edge approaches them and minute local fractures occur.

The graphite in cast iron and particles of a hard compound in bronze are examples of the presence of secondary stress raisers, which improve machinability. Elements may be added to alloys, other than those mentioned above, to improve machinability. These include manganese, molybdenum, zirconium, sulphur, carbon and selenium.

Let us now look briefly at one or two methods of joining materials, in particular welding and soldering. In welding brazing and soldering fusion takes place at the surfaces of the metals being joined. Soldering and brazing are fundamentally similar processes in that the joining material always melts at a temperature that is lower than the work piece.

Soldering or *soft* soldering as it is often called can be described as a process in which temperatures below 450°C are involved, whereas brazing temperatures are generally between 600 and 900°C. Welding can be achieved by hammering the surfaces in contact together at high temperatures, so that crystal growth takes place.

When *soldering* the solder must be capable of spreading across the surfaces of the metals to be joined, the solder is often assisted in this process by use of a *flux* or *wetting agent*. Solders produced from tin–lead alloys have good flow properties and melt at temperatures (183 to 250°C) that are comfortably below the temperatures of the metals to be joined.

In *fusion welding* processes either thermo-chemical sources, electric arc or some form of radiant energy can be used to melt the weld metal. In gas welding the surfaces to be joined are melted by a flame from a gas torch, the gases most commonly used being suitable mixtures of oxygen and acetylene.

In gas welding and other fusion welding processes a welding rod, is used to supply the necessary metal for the weld (the weld joint being suitably prepared prior to the commencement of welding). Other fusion welding processes apart from gas welding include, electric arc where the arc is struck between a carbon electrode and the work itself.

Solid phase welding processes include smith welding (described earlier), ultrasonic welding, friction welding and diffusion welding. I will just mention the diffusion welding process. In diffusion welding the sheets to be joined are held together under light pressure in a vacuum chamber. The temperature is then raised sufficiently for diffusion to occur across the interface so that the surfaces become joined by a region of solid solution. Steel can be clad with brass in this way.

Some of the other joining processes such as rivetting, and joining by adhesives I have asked you to investigate as an exercise at the end of this element. I do, however, consider it necessary to say something about the advantages and disadvantages of using *adhesives* to join materials.

Possibly the principal advantage of an adhesive bond is that the adhesive fastens to the entire bonded surface, thereby distributing the load more evenly and thus avoiding high localized stresses. Adhesives may also be used to advantage instead of riveting. Rivets add weight to structures, increase the likelihood of corrosion and can look unsightly on many domestic products. Materials with different coefficients of expansion can be joined by elastomer type adhesives that take the strain at the adhesive joint rather than the materials being joined. Adhesives can also be cured at relatively low temperatures, so preventing any unnecessary damaged to the materials being joined (adherends).

Adhesive joints also have one or two disadvantages primarily they are very much restricted in use at high temperatures. Also components cannot easily be dismantled for maintenance. Finally surface cleanliness and process control are very important, this usually requires a considerable amount of equipment.

Polymer adhesives are by far the most commonly used for engineering and industrial applications.

So much then for the processing of metals and metal alloys, let us now focus our attention on the processing of polymers.

Non-metals

Polymer processing

An important attraction of polymer compounds is that they can be readily converted into a variety of useful shapes. Polymer processing is concerned with the technology needed to make articles from polymer compounds. Three common themes underly most of the methods. The first involves making the appropriate compound, usually in liquid form, from the raw ingredients. The second is concerned with transforming the compound into useful shapes. Finally the third theme is to ensure that once the product has been formed it retains its shape and dimensions.

So the first theme is concerned with mixing the raw ingredients to produce the correct compound. There are two basic stages in the mixing process, dispersing the additives in the polymer and achieving a uniform shapeable state. The two main processes used for good dispersion of the polymers ingredients, particularly in the rubber industry, are the two-roll mill and the intensive mixer. In the intensive mixer two rotors counter-rotate within a robust casing. There is only a small clearance between the tips of the rotors and the casing. Ingredients are fed in through an opening in the top of the machine, which may be closed during use. The fully mixed compound being removed through a port in the underside of the casing.

The two-roll mill consists of two heavy horizontal cylindrical polished rollers fitted with water-cooling channels (Figure 3.49). The rollers counter rotate and are separated by a narrow gap. The feed is dropped between the two rollers. The sticky rubber substance is dragged down between the two rollers and adheres to them in the form of a band. This band is cut on a helix at an angle to the roll-axis and it is simply folded over.

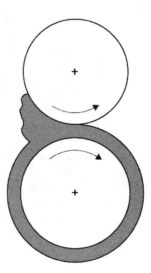

Figure 3.49 *Polymer being processed in a two-roll mill*

Because of the high viscosity of the rubber compound, the power to drive the rollers is high, but because the water–cooled rollers have a large surface area in contact with the thin band of polymer, this prevents excessive temperature rise. It is therefore safe to add the vulcanizing (cross-linking) agents to the mix.

Extrusion processes are used to make continuous products of constant cross-section, usually from thermoplastics or rubber compounds. Two basic types of equipment are commonly used these are the *calender* and the *screw extruder*.

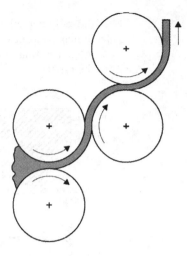

Figure 3.50 *A typical calendering process for a polymer*

The *calender* consists of four heavy rollers that are sometimes known as *bowls* (Figure 3.50). The top two bowls act just like the two-roll mill previously described. The compound is drawn through the first two rollers and adheres to them. It is thus transferred to the next pair where its cross-section is further reduced. Finally the required surface finish, either plain or embossed is supplied as the polymer mix passes through the final stage rollers. Sheet up to a few millimetres thick and a metre or so wide can be produced in this way.

The *screw extruder* consists of one or sometimes two screws that rotate inside a close-fitting barrel of constant diameter. The screw is driven by a large electric motor, being connected to the screw by a reduction gearbox. The screw has to deliver a steady stream of molten polymer to the die. The screw acts as a pump, which develops drag flow and thus heat which helps to melt or plasticize the feed. A hopper is used to supply the feed to the screw. A typical screw extruder is shown in Figure 3.51.

Compression moulding of polymers is similar to cold pressing metals. In that a thermosetting polymer compound known as the charge is fed into the jaws of a moulding die. Pressure is applied and

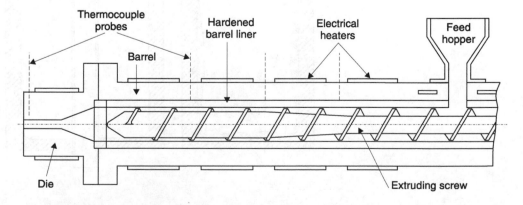

Figure 3.51 *A typical polymer screw extruder*

the polymer takes up exactly the shape of the die to form a three dimensional product. The stages of this process are shown in Figure 3.52. The final product is ejected from the mould after completion of the process.

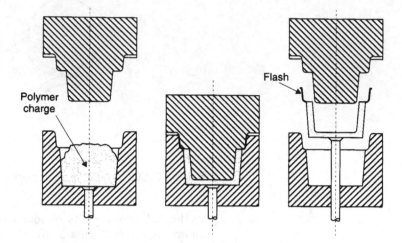

Figure 3.52 *Polymer compression moulding process*

In *extrusion blow moulding* a thick walled thermoplastic called a *parison* is extruded vertically downwards between the open faces of a cold split mould, which produces a hollow cavity. The mould is then closed and sealed and the still warm parison is inflated with compressed air so that the outside of the tube takes up the shape of the mould (Figure 3.53). Plastic bottles and other useful containers are often made in this way.

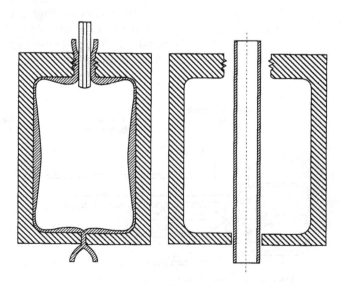

Figure 3.53 *Extrusion blow moulding*

Let us consider one final moulding process that of *injection moulding.* In this process polymer melt is injected into an impression within a closed split mould which has the dimensions required for the finished product, see Figure 3.54.

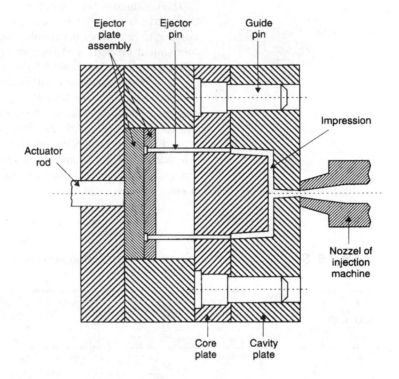

Figure 3.54 *Typical injection moulder*

Thermoplastics, thermosets and elastomers can all be injection moulded. Many aspects of the construction of compression and injection moulds are similar in principle.

We could consider many other pieces of polymer processing equipment, such as hand lay-up techniques, but these are perhaps better covered in the next section, when I deal with composite materials.

Composite materials and their processing

This range of materials has been left to last because *composites* are a mixture of two or more constituents or phases, rather than a distinct class. I have already introduced you to composites when we talked initially about classes of material. However that definition was not really sufficient and three other criteria have to be satisfied in order to classify a material as a composite. First both constituents have to be present in reasonable proportions, rather than fractions of a per cent. Second, the constituent phases will have different properties, so that the properties of the resulting composite are different from its component parts. Finally, a composite is usually produced by intimately mixing and combining the constituents by various means.

So, for example, a metal alloy which has a two-phase microstructure which is produced during solidification from the melt, or by subsequent heat treatment, has not involved intimate mixing and cannot therefore, be classified as a true composite material.

The continuous constituent of a composite is known as the *matrix*, the matrix is often but not always present in the greater quantity. A composite may have a ceramic, metallic or polymeric matrix. The mechanical properties of composites produced from these matrices differ completely, as Table 3.6 shows.

The second constituent within the composite is referred to as the *reinforcing phase*, because this phase strengthens or enhances in some way the properties of the matrix. The reinforcement may take the form of small particles, chopped strand, continuous strand or woven matting. Dependent on how it is mixed, will generally dictate whether or not the composite has unidirectional or multi-directional strength and stiffness characteristics. Figure 3.55 illustrates this point.

Table 3.6 *Some properties of ceramics, metals and polymers*

Material	Density (Mg/m³)	Young's Modulus (Gpa)	Strength (Mpa)	Ductility (%)	Toughness K_{IC} (Mpa m^½)	Specific modulus (GPa)/ (Mg/m³)	Specific strength (MPa)/ (Mg/m³)
Ceramics							
Alumina	3.87	382	332	0	4.9	99	86
Magnesia	3.60	207	230	0	1.2	58	64
Silicon nitride		166	210	0	4.0		
Zirconia	5.92	170	900	0	8.6	29	152
β-Sialon	3.25	300	945	0	7.7	92	291
Metals							
Aluminium	2.70	69	77	47	~30	26	29
Aluminium alloy	2.83	72	325	18	~25–30	25	115
Brass	8.50	100	550	70	–	12	65
Nickel alloy	8.18	204	1200	26	~50–80	25	147
Steel mild	7.86	210	460	35	~50	27	59
Titanium alloy	4.56	112	792	20	~55–90	24	174
Polymers							
Epoxy	1.12	4	50	4	1.5	4	36
Nylon 6.6	1.14	2	70	60	3–4	18	61
Polyetheretherketone	1.30	4	70		1.0	3	54
Polymethylmethacrylate	1.19	3	50	3	1.5	3	42
Polystyrene	1.05	3	50	2	1.0	3	48
Polyvinylchloride rigid	1.70	3	60	15	4.0	2	35

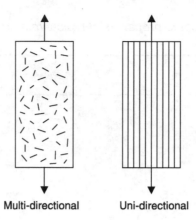

Multi-directional   Uni-directional

Figure 3.55 *Reinforcing fibre orientation dictates whether or not the composite material has multi-directional or uni-directional properties*

In this text I will be concentrating on just one major group of composites *polymer matrix composites* (PMCs). This group has a wide range of engineering applications and by studying them, the general principles underlying their properties and processing can be extended to other groups of composite such as those with a ceramic or metal matrix. PMCs use all three classes of polymer; thermoplastics, thermosets and elastomers, although thermosetting polymers dominate the market for this type of composite.

Carbon fibre reinforced plastic (CFRP) is a very well known polymer matrix composite, where the reinforcement is carbon fibre. Other reinforcing materials used with PMCs include glass, polyethylene, boron, and Kevlar (a type of aramatic polyamide). A range of reinforcing fibres and matrices, with their mechanical properties is given in Table 3.7.

Let us now concentrate on one or two processing methods for PMCs. In *hand lay-up* the reinforcement is put down to line a mould previously treated with some form of release agent to prevent sticking. The reinforcement can be in many forms, such as chopped strand mat, woven mat, etc. The liquid thermosetting resin is mixed with a curing (setting) agent and applied with a brush or roller, ensuring that the resin is thoroughly worked into the reinforcement. The most commonly used resins are polyesters and curing normally takes place at room temperature. Hand lay-up is labour intensive but requires little capital equipment. For these reasons it is often used to produce one-off specialist articles or large components such as swimming pools and boat hulls.

Another manual method of production is known as *spray-up*. In this method a spray gun charged with the matrix resin, chopped fibres, and curing agent, is used. This method is quick, cheap and efficient, although at suitable intervals the sprayed composite has to be rolled to release trapped air.

Let us consider just two moulding methods for the production of composite components, *die-moulding* and *bag-moulding* Die-moulding is widely used for long production runs for components ranging in size from small domestic items to large commercial vehicle panels. The material to be shaped is pressed between heated

Table 3.7 *Some properties of typical reinforcing fibres and matrix materials*

Material	Relative density	Young's modulus (GPa)	Tensile strength (GPa)
Reinforcing fibres			
E Glass	2.55	72	1.5–3.0
S Glass	2.5	87	3.5
Carbon-pitch	2.0	380	3.0-3.6
Carbon-pan	1.8	220–240	2.3–3.6
SiC whisker	3.2	480	7.0
Kevlar	1.47	130–180	2.6–3.5
Nomex	1.4	17.5	0.7
Polyamide (typical)	1.4	5.0	0.9
Matrix materials			
Steel	7.8	210	0.34–2.1
Aluminium alloys	2.7	70	0.14–0.62
Epoxy resin	1.2	2–3.5	0.05–0.09
Polyester resin	1.4	2–3.0	0.04–0.08

matched dies. The pressure required may be as high as 50 MPa. The feed material flows into the contours of the mould and when the temperature is high enough it rapidly cures. Good dimensional accuracy and detail are possible with this method, depending on the quality of the die used. The feed material, which already contains all necessary ingredients and the curing agent, may be in the form of a sheet or simply fed in as a dough.

In the *bag-moulding* process only one half of the mould is used to shape the component in *vacuum-bagging* the laid-up material, which consists of heated pre-impregnated reinforcement, is sealed by a bag over the component. When vacuum pressure is applied to the bag, sufficient pressure is applied to the work piece to cure it (Figure 3.56).

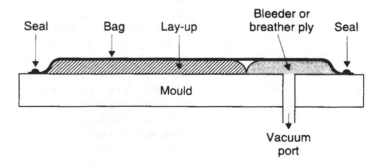

Figure 3.56 *Typical lay-up procedure using the vacuum-bagging process*

Autoclave moulding is a modification of vacuum forming where pressure in excess of atmospheric is used to produce high density products for critical applications, such as racing car and aircraft components. An *autoclave* is a re-circulating oven, that is pressurized by an inert gas, often nitrogen. The sealed bag is still used to stabilize the component against the mould (Figure 3.57).

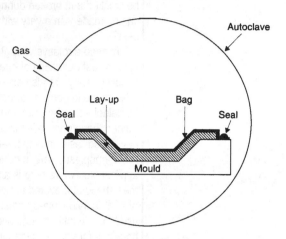

Figure 3.57 *Pressurized autoclave moulding process*

Moulding techniques for PMCs concludes our look at materials processing, there are many methods that I have not mentioned but space prevents me from doing so. The assignment at the end of this unit, asks you to investigate some of these processes.

Selecting materials for engineered products

In this final element we will attempt to select materials for a variety of engineering applications. On what criteria must we base our assumptions?

When you are faced with making choices for suitable materials you should consider their: static strength and stiffness, structural efficiency, fatigue resistance, corrosion resistance, as well as their availability, cost and fabrication characteristics. For specialized requirements you may also have to consider their: erosion, abrasion and wear characteristics; their compatibility with other materials, and their thermal and electrical properties.

The following examples may help to provide the answers:

Example 3.6

Select suitable candidate materials for the manufacture of a domestic three-pin electrical plug.

We need to first familiarize ourselves with the components that go to make up a plug. There is the plug body, electrical lead connectors, fuse holder, and the three conducting pins.

The plug body will need to be an excellent insulator, if we are to handle

---

the plug safely. It will require some resistance to the heat that may be generated by the electrical wiring, due to the power being carried. It needs to be reasonably tough and resistant to rough handling. The body also needs to be manufactured from a material that can easily be mass produced at a competitive price.

Ceramics are excellent insulators, but are brittle and fragile, so can easily be cracked and broken during use. Polymers are excellent insulators and can be made with a very wide range of properties. So which polymers are best?

We need a relatively tough and resilient material. Rubbers are able to absorb vast quantities of energy and therefore would be resistant to rough handling. They are also excellent insulators. The melting temperature of rubbers are in general lower than thermoset polymers such as phenol-formaldehyde (phenolic) and melamine-formaldehyde (melamine). Natural rubbers have poor resistance to the environment, although butyl-rubbers have good resistance to heat and chemicals.

Thermoplastics are in general not suitable for plug bodies and similar applications where heat is a problem, because they tend to soften and lose their shape at elevated temperatures. Thermoplastics, however, in the form of ABS (acrylonitrile-butadiene-styrene) and PVC (polyvinyl chloride) are used for a wealth of domestic applications such as telephone hand-sets, housings for vacuum cleaners and grass mowers, water pipes and guttering, insulation for wire and curtain rails.

The decision on which polymer to chose, is not quite as simple as you may have first thought. From the above discussion, a thermoset plastic appears to have the desired properties. In fact phenolics are used predominantly for the manufacture of plugs, switches, sockets and car distributor heads. Various rubbers are often used for plugs and electrical connectors that are required for outdoor applications.

The fuse holder link, terminal screws and conducting pins, all need to be made from a good conducting material. In addition the terminal screws and fittings need to be robust enough to take the torsional loads created by fitting and removing the conducting wires from the appliance.

Suitable conductors might include, aluminium, silver and copper. Silver is a relatively soft and weak material and is expensive, so can be eliminated, even though its conductivity is slightly better than copper.

The conductivity of both copper and aluminium is excellent, but pure aluminium is weaker than copper, so perhaps we should choose copper. However, copper tends to oxidize (passivate) very easily, particularly at elevated temperatures, which might be experienced at the terminals. So pure copper might not be totally appropriate, if we alloy copper with zinc, we produce brass, which becomes much less susceptible to oxidation, but retains the strength characteristics necessary at the terminals.

Brass is in fact commonly used for the terminal fittings and screws of electrical plugs, as well as the conducting pins. The fuse links may be pure copper or brass.

All polymer materials can be easily moulded and are very suitable for mass production. Copper, zinc and aluminium are also easy to produce and for the small amounts required in the conducting parts of a plug, they are relatively inexpensive.

Finally, note that in our choice of materials for an electrical plug, there are many similar candidate materials that may be used. It is the job of the materials engineer to make the best choice for each and every engineering application.

For our final example, we will consider materials selection in a slightly different way:

Example 3.7

Compare and contrast the differences in the materials and fabrication techniques necessary for the manufacture of:

(a) a large industrial flywheel used for a crank press, and
(b) a flywheel for a child's clockwork toy.

First we need to consider what a flywheel is and what it is used for.

A flywheel is designed to store energy and use up this stored energy when required. This is achieved by winding up a large mass and rotating it at a suitable speed. The rotational kinetic energy produced at the shaft enables further rotation to take place even after the power source has been removed.

In the crank press the flywheel is run up using an electrical motor, the stored energy is then used for the actual pressing operation itself. This avoids the need for a very heavy duty and powerful motor, which would be expensive and less energy efficient. So the flywheel needs to be heavy, tough and resistant to fatigue and creep loads. It must also have sufficient strength to avoid *bursting* at high rotational velocities. Steel alloys and cast irons are often used for flywheel construction.

The size of flywheel necessary for typical pressing operations may be up to two metres in diameter possibly weighing two tons or more, dependent obviously on the speed of operation and the energy required for pressing operations.

If we are interested in one-off or very small batch production, we could sand cast the flywheel, and machine the hub to ensure that the flywheel runs true. However machining operations are time consuming and expensive, especially for such a large structure.

An alternative might be to fabricate the flywheel from smaller segmented parts and bolt these together. Another alternative might be to spin the outer, very large rim of the flywheel and weld to the inner flywheel disk and pre-turned flywheel hub. By virtue of the size of the flywheel, it is unlikely that a die casting or forging process would be used.

Now how does this compare with the flywheel used in a child's clockwork toy? Well obviously there is the question of scale, the toy flywheel will be very small, able to supply sufficient energy to propel the toy once released, but not too large to prevent the child winding up the toy.

Safety must also be considered, the flywheel should not have sharp edges or be manufactured from toxic materials. Metals in this case may not be appropriate; the material will still need to be tough and robust but will not need to be as dense as that used for the industrial version. It will also need to be made from a material capable of mass production.

The loads imposed on the flywheel will be minimal and elevated temperatures are not involved. Thus an ideal candidate might be a material from the polymer range. Thermoplastics such as polyethylene, polypropylene and nylon would be suitable. All these polymers can be injection moulded, making them eminently suitable for large scale production.

In order to make choices about suitable materials, there are many sources of reference to help you. For example, data on the properties of materials can be found in reference books and technical data books, held in most libraries. Specific information about components and specific materials may be obtained from manufacturers data catalogues. Information on materials testing procedures can be obtained from British Standards Publications.

Finally then, try to remember that, when selecting a material, a balance has to be reached between its mechanical and physical properties, its ease of fabrication and availability, as well as its cost!

Activity 3.9

A toy manufacturing company, Micro Models, is about to embark on the manufacture of a working model (1:12) traction engine. They have asked you to advise them on the selection of materials to be used in various component parts of the model traction engine. Suggest, with reasons, materials to be used for each of the following:

(a) engine flywheel
(b) driving wheels
(c) chassis
(d) steam cylinder
(e) water tank
(f) mechanical linkages.

Present your recommendations in the form of a briefing pack for the Technical Director and Production Manager of Micro Models. Your briefing pack should be word processed and should contain relevant data and specifications, where appropriate.

Review questions

1   Describe, briefly, the structure of a pure metal.

2   Explain what is meant by the term *composite material*. Give THREE examples of common composite materials.

3   Name, and briefly describe the properties of, THREE main types of *polymer material*.

4   Name, and briefly describe the properties of, THREE *ceramic materials*.

5   Classify each of the materials listed below as either metals, polymers, ceramics or composite materials:

(a)   aluminium
(b)   clay
(c)   polyvinyl chloride (PVC)
(d)   rubber
(e)   glass
(f)   tungsten
(g)   brick
(h)   wood.

6   Sketch graphs showing how the electrical resistance of each of the following types of material varies with temperature:

(a)   conductors
(b)   insulators
(c)   semiconductors.

7   Briefly explain each of the following terms in relation to the mechanical properties of materials:

(a)   ductility
(b)   malleability
(c)   plasticity
(d)   hardness
(e)   toughness.

8   State *Hooke's Law*.

9   Define *Young's modulus*.

10   Sketch typical stress–strain graphs for the following types of material:

(a)   a brittle polymer
(b)   a cold-drawn polymer
(c)   a ceramic material.

11   A steel reinforcing rod has a cross-sectional area of 240 mm^2. Determine the compressive stress on the rod if it is subject to a compressive force of 720 N.

12   In a tensile test a specimen is subjected to a strain of 0.05%. Determine the change in length of the specimen if it has an unstrained length of 250 mm.

13 A metal bar has a cross-sectional area of 350 mm^2. If the material has a yield stress of 225 MPa, determine the tensile force that must be applied in order to cause yielding.

14 Define *Poisson's ratio*.

15 Briefly explain each of the following terms in relation to the electrical properties of materials:

(a) resistivity
(b) conductivity.

16 State two examples of *semiconductor materials*.

17 Sketch a typical hysteresis curve for a ferromagnetic material. Label your drawing clearly.

18 Describe the essential properties of materials that will be used in each of the following electrical/electronic applications:

(a) the magnetic core of a transformer
(b) the field coil windings in a generator
(c) the dielectric between the plates of a capacitor
(d) the material used for fabricating an integrated circuit.

19 State TWO materials that have a low value of thermal conductivity and TWO materials that have a high value of thermal conductivity. Suggest a typical engineering application for each of these classes of material.

20 In relation to the structure of materials, explain each of the following terms:

(a) bonding
(b) valency.

21 Describe THREE types of corrosion that can affect metals. Explain how each type of corrosion is caused.

22 Describe how each of the following tests are carried out on samples of a material:

(a) hardness
(b) bending
(c) toughness.

Unit 4 Applied science in engineering

Summary

This section aims to develop in the reader an understanding of fundamental science concepts and to give a basic mechanical, thermal and electrical engineering systems background for student engineers.

More specifically, the aims are to describe engineering systems in terms of basic scientific laws and principles, to investigate the behaviour of simple linear systems in engineering, to calculate the response of engineering systems to changes in variables, and to determine the response of such engineering systems to changes in parameters.

SI units

The system of units used in engineering and science is the *Système Internationale d'Unités* (international system of units), usually abbreviated to SI units, and is based on the metric system. This was introduced in 1960 and is now adopted by the majority of countries as the official system of measurement.

The basic units in the SI system are listed below with their symbols:

Quantity	Unit and symbol
length	metre, m
mass	kilogram, kg
time	second, s
electric current	ampere, A
thermodynamic temperature	Kelvin, K
luminous intensity	candela, cd
amount of substance	mole, mol

SI units may be made larger or smaller by using *prefixes* that denote multiplication or division by a particular amount. The eight most common multiples, with their meaning, are listed below:

Prefix	*Name*	*Meaning*
T	tera	multiply by 1 000 000 000 000 (i.e. $\times 10^{12}$)
G	giga	multiply by 1 000 000 000 (i.e. $\times 10^{9}$)
M	mega	multiply by 1 000 000 (i.e. $\times 10^{6}$)
k	kilo	multiply by 1 000 (i.e. $\times 10^{3}$)
m	milli	divide by 1 000 (i.e. $\times 10^{-3}$)
μ	micro	divide by 1 000 000 (i.e. $\times 10^{-6}$)
n	nano	divide by 1 000 000 000 (i.e. $\times 10^{-9}$)
p	pico	divide by 1 000 000 000 000 (i.e. $\times 10^{-12}$)

Fundamental concepts

Length, area, volume and mass

Length is the distance between two points. The standard unit of length is the *metre* (m) although the *centimetre* (cm), *millimetre* (mm) and *kilometre* (km) are often used.

1 cm = 10 mm; 1 m = 100 cm = 1000 mm; 1 km = 1000 m

Area is a measure of the size or extent of a plane surface and is measured by multiplying a length by a length. If the lengths are in metres then the unit of area is the square metre, m^2.

$1 \ m^2 = 1 \ m \times 1 \ m = 100 \ cm \times 100 \ cm = 10 \ 000 \ cm^2$ or $10^4 \ cm^2$

$1 \ m^2 = 1 \ 000 \ mm \times 1 \ 000 \ mm = 1 \ 000 \ 000 \ mm^2$ or $10^6 \ mm^2$

Conversely $1 \ cm^2 = 10^{-4} \ m^2$ and $1 \ mm^2 = 10^{-6} \ m^2$

Volume is a measure of the space occupied by a solid and is measured by multiplying a length by a length by a length. If the lengths are in metres then the unit of volume is in cubic metres, m^3.

$1 \ m^3 = 1 \ m \times 1 \ m \times 1 \ m$

$= 100 \ cm \times 100 \ cm \times 100 \ cm = 10^6 \ cm^3$

$= 1000 \ mm \times 1000 \ mm \times 1000 \ mm$

$= 10^9 \ mm^3$

Conversely, $1 \ cm^3 = 10^{-6} \ m^3$ and $1 \ mm^3 = 10^{-9} \ m^3$

Another unit used to measure volume, particularly with fluids, is the litre, l, where $1 \, l = 1000 \, cm^3$.

Mass is the amount of matter in a body and is measured in *kilograms*, kg.

$1 \, kg = 1000 \, g$ (or conversely, $1 \, g = 10^{-3} \, kg$)

and

1 tonne (t) $= 1000 \, kg$

Example 4.1

Express (a) a length of 36 mm in metres, (b) 32 400 mm² in square metres, and (c) 8 540 000 mm³ in cubic metres.

(a) $1 \, m = 10^3 \, mm$ or $1 \, mm = 10^{-3} \, m$

Hence $36 \, mm = 36 \times 10^{-3} \, m = \dfrac{36}{10^3} \, m = 0.036 \, m$

(b) $1 \, m^2 = 10^6 \, mm^2$ or $1 \, mm^2 = 10^{-6} \, m^2$

Hence $32 \, 400 \, mm^2 = 32400 \times 10^{-6} \, m^2 = \dfrac{32400}{10^6} \, m = 0.0324 \, m^2$

(c) $1 \, m^3 = 10^9 \, mm^3$ or $1 \, mm^3 = 10^{-9} \, m^3$

Hence $8 \, 540 \, 000 \, mm^3 = 8 \, 540 \, 000 \times 10^{-9} \, m^3$

$$= \dfrac{8540000}{10^9} \, m = 8.54 \times 10^{-3} \, m^3 \text{ or } 0.00854 \, m^3$$

Example 4.2

A cube has sides each of length 50 mm. Determine the volume of the cube in cubic metres.

Volume of cube $= 50 \, mm \times 50 \, mm \times 50 \, mm = 125 \, 000 \, mm^3$

$1 \, mm^3 = 10^{-9} \, m$, thus volume $= 125 \, 000 \times 10^{-9} \, m^3$

$$= 0.125 \times 10^{-3} \, m^3$$

Test your knowledge 4.1

1 Determine the area of a room 15 m long by 8 m wide in (a) m², (b) cm² and (c) mm²

2 A bottle contains 4 litres of liquid. Determine the volume in (a) cm³, (b) m³ and (c) mm²

Density

Density is the mass per unit volume of a substance. The symbol used for density is ρ (Greek letter rho) and its units are kg/m^3.

$$\text{Density} = \frac{\text{mass}}{\text{volume}} = \frac{m}{V} \text{ or } m = V\rho \text{ or } V = \frac{m}{\rho}$$

where m is the mass in kg, V is the volume in m^3 and ρ is the density in kg/m^3

Some typical values of densities include:

Aluminium	2700 kg/m^3	Steel	7800 kg/m^3
Cast iron	7000 kg/m^3	Petrol	700 kg/m^3
Cork	2500 kg/m^3	Copper	8900 kg/m^3
Lead	11400 kg/m^3	Water	1000 kg/m^3

Example 4.3

Determine the density of 50 cm^3 of copper if its mass is 445 g.

Volume = 50 cm^3 = 50 $\times$ 10^{-6} m^3; mass = 445 g = 445 $\times$ 10^{-3} kg

$$\text{Density} = \frac{\text{mass}}{\text{volume}} = \frac{445 \times 10^{-3}}{50 \times 10^{-6}} = 8.9 \times 10^3 \text{ kg/m}^3 \text{ or } \mathbf{8900 \text{ kg/m}^3}$$

Test your knowledge 4.2

1 Determine the density of 80 cm^3 of cast iron if its mass is 560 g.

2 Determine the volume, in litres, of 35 kg of petrol of density 700 kg/m^3.

3 A piece of metal 200 mm long, 150 mm wide and 10 mm thick has a mass of 2700 g. What is the density of the metal?

Example 4.4

The density of aluminium is 2700 kg/m^3. Calculate the mass of a block of aluminium if it has a volume of 100 cm^2.

Density, ρ = 2700 kg/m^3; volume V = 100 cm^3 = 100 $\times$ 10^{-6} m^3

Since density = mass/volume, then mass = density $\times$ volume.

Hence $m = \rho V$ = 2700 kg/m^3 $\times$ 100 $\times$ 10^{-6} m^3 kg = 0.270 kg or **270 g**

Scalar and vector quantities

Quantities used in engineering and science can be divided into two groups:

(a) *Scalar quantities* have a size (or magnitude) only and need no other information to specify them. Thus, 10 centimetres, 50 seconds, 7 litres and 3 kilograms are all examples of scalar quantities.

(b) *Vector quantities* have both a size or magnitude and a direction, called the line of action of the quantity. Thus, a velocity of 50 kilometres per hour due east, an acceleration of 9.81 metres per second squared vertically downwards and a force of 15 newtons at an angle of 30 degrees are all examples of vector quantities.

Force and work

When forces are all acting in the same plane, they are called *coplanar*. When forces act at the same time and at the same point, they are called *concurrent* forces.

Force is a *vector quantity* and thus has both a *magnitude* and a *direction*. A *vector* can be represented graphically by a line drawn to scale in the direction of the line of action of the force

If a body moves as a result of a force being applied to it, the force is said to do work on the body. The amount of work done is the product of the applied force and the distance, i.e.

work done = force × distance moved in the direction of the force

The unit of work is the *joule* (J) which is defined as the amount of work done when a force of 1 newton acts for a distance of 1 m in the direction of the force. Thus

$1\text{ J} = 1\text{ N m}$

If a graph is plotted of experimental values of force (on the vertical axis) against distance moved (on the horizontal axis) a force/distance graph or work diagram is produced. *The area under the graph represents the work done.*

For example, a constant force of 20 N used to raise a load a height of 8 m may be represented on a force/distance graph as shown in Figure 4.1(a). The area under the graph, shown shaded, represents the work done. Hence

work done = 20 N × 8 m = 160 J

Similarly, a spring extended by 20 mm by a force of 500 N may be represented by the work diagram shown in Figure 4.1(b).

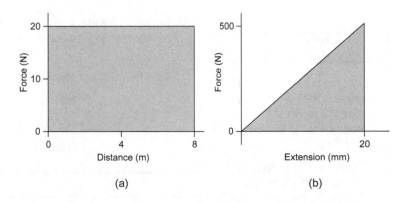

(a)   (b)

Figure 4.1 *Force plotted against (a) distance and (b) extension*

work done = shaded area

$= \frac{1}{2} \times \text{base} \times \text{height}$

$= \frac{1}{2} \times (20 \times 10^{-3})\text{ m} \times 500\text{ N} = \mathbf{5\ J}$

Example 4.5

Calculate the work done when a force of 40 N pushes an object a distance of 500 m in the same direction as the force.

work done = force × distance moved in the direction of the force

= 40 N × 500 m = 20000 J (since 1 J = 1 N m)

i.e. work done = **20 kJ**

Example 4.6

Calculate the work done when a force of 40 N pushes an object a distance of 500 m in the same direction as the force.

A motor supplies a constant force of 1 kN which is used to move a load a distance of 5 m. The force is then changed to a constant 500 N and the load is moved a further 15 m. Draw the force/distance graph for the operation and from the graph determine the work done by the motor.

The force/distance graph of work diagram is shown in Figure 4.2.

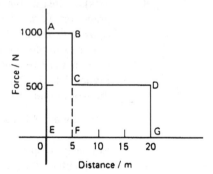

Figure 4.2

Between points A and B a constant force of 1000 N moves the load 5 m; between points C and D a constant force of 500 N moves the load from 5 m to 20 m.

Total work done = area under the force/distance graph

= area ABFE + area CDGF

= (1000 N × 5 m) + (500 N × 15 m)

= 5000 J + 7500 J = 12500 J = **12.5 kJ**

Example 4.7

Calculate the work done when a mass of 20 kg is lifted vertically through a distance of 5.0 m.

The force to be overcome when lifting a mass of 20 kg vertically upwards is $(m \times g)$, i.e. 20 × 9.81 = 196.2 N

Work done = force × distance = 196.2 × 5.0 = **981 J**

1 Calculate the work done when a mass is lifted vertically by a crane to a height of 5 m, the force required to lift the mass being 98 N.

2 A spring requires a force of 50 N to cause an extension of 100 mm. Determine the work done in extending the spring (a) from 0 to 100 mm, and (b) from 40 mm to 100 mm.

3 Calculate the work done when a mass of 50 kg is lifted vertically through a distance of 30 m.

Energy applications

Types of energy

Energy is the capacity, or ability, to do work. The unit of energy is the joule, the same as for work. Energy is expended when work is done. There are several forms of energy and these include:

- Mechanical energy
- Heat or thermal energy
- Electrical energy
- Chemical energy
- Nuclear energy
- Light energy
- Sound energy.

Energy may be converted from one form to another. The *principle of conservation of energy* states that the total amount of energy remains the same in such conversions, i.e. energy cannot be created or destroyed. An example of energy conversion is the conversion of mechanical energy to electrical energy by a generator.

Activity 4.1

Energy is often converted from one form to another as part of the normal operation of an engineering system. For EACH of the engineering systems listed below, identify the energy conversions that take place and explain how each conversion is achieved. Present your findings in the form of a brief word processed report.

1 The engine and transmission used in a small car.

2 An emergency lighting system that uses a rechargeable battery.

3 A wind generator that supplies power to a remote rural community.

Efficiency

Efficiency is defined as the ratio of the useful output energy to the input energy. The symbol for efficiency is η. Hence

$$\text{efficiency, } \eta = \frac{\text{useful output energy}}{\text{input energy}}$$

Efficiency has no units and is often stated as a percentage. A perfect machine would have an efficiency of 100%. However, all machines have an efficiency that is lower than this due to friction and other losses. Thus, if the input energy to a motor is 1000 J and the output energy is 800 J then the efficiency is

$$\frac{800}{1000} \times 100\%, \quad \text{i.e. } 80\%$$

Example 4.8

A machine exerts a force of 200 N in lifting a mass through a height of 6 m. If 2 kJ of energy is supplied to it, what is the efficiency of the machine?

Work done in lifting mass   = force × distance moved

= weight of body × distance moved

= 200 N × 6 m = 1200 J = useful energy output

Energy output = 2 kJ = 2000 J

$$\text{Efficiency} = \frac{\text{useful output energy}}{\text{input energy}} = \frac{1200}{2000} = \textbf{0.6 or 60\%}$$

Activity 4.2

Engineers are often concerned with looking for ways of improving the efficiency of an engineering system. With reference to an engineering system of your choice, explain why efficiency is important and describe practical measures that have been taken to improve it. Present your answer in the form of a brief word processed report.

Activity 4.3

A small engineering company, Ace Electronics, has asked you to investigate methods that could be used to improve the energy efficiency of their manufacturing plant. The Technical Director has asked you to look for suitable materials that can be used as insulation within the roof space of the building. You have been asked to present your findings in the form of a word processed data sheet on each material. The data sheet should provide appropriate technical specifications on each material including its composition, thermal characteristics, installation technique, and any handling precautions. You should also information on manufacturers and relative costs.

Example 4.9

A hoist has an efficiency of 80%. Determine the input energy required if the hoist exerts a force of 450 N when lifting a load through a distance of 4 m.

Work done in lifting mass = force × distance moved

= weight of body × distance moved

= 450 N × 4 m = 1800 J

$$\text{Input energy} = \frac{\text{useful output energy}}{\text{efficiency}}$$

Hence

$$\text{Input energy} = \frac{1800}{0.8}$$

= 2250 J = **2.25 kJ**

Example 4.10

A motor drive exerts a force of 15 N when moving a component through a distance of 0.7 m. If the input energy to the motor drive is 20 J determine the efficiency of the motor drive. Also calculate the input energy to the motor drive if the efficiency of the system increases by 10%.

Work done by motor drive = force × distance moved

= 150 N × 0.7 m = 10.5 J

Input energy = 20 J

Now

$$\text{Efficiency} = \frac{\text{output energy}}{\text{input energy}}$$

Thus

$$\text{Efficiency} = \frac{10.5}{20} = 0.525 = \textbf{52.5\%}$$

If efficiency increases by 10%, the new efficiency will be

(52.5 + 10) = 62.5 % or 0.625

But

$$\text{Input energy} = \frac{\text{useful output energy}}{\text{efficiency}}$$

Hence

$$\text{Input energy} = \frac{10.5}{0.625}$$

= **16.8 J**

Example 4.11

4 kJ of energy are supplied to a machine used for lifting a mass. The force required is 800 N. If the machine has an efficiency of 50%, to what height will it lift the mass?

We can rearrange the formula, efficiency $= \dfrac{\text{useful output energy}}{\text{input energy}}$ to make

useful output energy the subject, as follows:

$$\text{useful output energy} = \text{efficiency} \times \text{input energy}$$

$$\text{Thus useful output energy} = \text{efficiency} \times \text{input energy}$$

$$= \frac{50}{100} \times 4000 \text{ J} = 2000 \text{ J}$$

Work done = force × distance moved,
hence 2000 J = 800 N × height,

$$\text{from which, height} = \frac{2000 \text{ J}}{800 \text{ N}} = \textbf{2.5 m}$$

Example 4.12

A hoist exerts a force of 500 N in raising a load through a height of 20 m. The efficiency of the hoist gears is 75% and the efficiency of the motor is 80%. Calculate the input energy to the hoist.

The hoist system is shown diagrammatically in Figure 4.3.

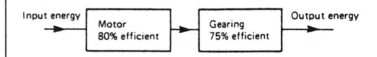

Figure 4.3

Output energy = work done = force × distance = 500 N × 20 m = 10000 J

For the gearing,

$$\text{efficiency} = \frac{\text{output energy}}{\text{input energy}}$$

i.e. $\dfrac{75}{100} = \dfrac{10000}{\text{input energy}}$

from which, the input energy to the gears = 10000 × (100/75) = 13333 J.

The input energy to the gears is the same as the output energy of the motor.

Thus, motor efficiency $= \dfrac{\text{output energy}}{\text{input energy}} = \dfrac{80}{100} = \dfrac{13333}{\text{input energy}}$

Input energy to the system = 13333 × 100/80 = 16670 J = **16.67 kJ**

Power

Power is a measure of the rate at which work is done or at which energy is converted from one form to another.

$$\text{Power, } P = \frac{\text{energy used}}{\text{time taken}} = \frac{\text{work done}}{\text{time taken}}$$

The unit of power is the watt (W) where 1 watt is equal to 1 joule per second. The watt is a small unit for many purposes and a larger unit called the kilowatt (kW) is used, where 1 kW = 1000 W. The power output of a motor which does 120 kJ of work in 30 s is thus given by

$$P = \frac{120 \text{ kJ}}{30 \text{ s}} = 4 \text{ kW}$$

Since work done = force × distance, then

$$\text{Power} = \frac{\text{work done}}{\text{time taken}} = \frac{\text{force} \times \text{distance}}{\text{time taken}}$$

$$= \text{force} \times \frac{\text{distance}}{\text{time taken}}$$

However,

$$\frac{\text{distance}}{\text{time taken}} = \text{velocity}$$

Hence

$$\text{power} = \text{force} \times \text{velocity}$$

Example 4.13

The output power of a motor is 8 kW. How much work does it do in 30 s?

Power = (work done)/(time taken), from which,

work done = power × time = 8000 W × 30 s = 240000 J = **240 kJ**

Example 4.14

Calculate the power required to lift a mass through a height of 10 m in 20 s if the force required is 3924 N.

Work done = force × distance moved = 3924 N × 10 m = 39240 J

$$\text{Power} = \frac{\text{work done}}{\text{time taken}} = \frac{39240 \text{ J}}{20 \text{ s}} = 1962 \text{ W or } \textbf{1.962 kW}$$

Example 4.15

A car hauls a trailer at 90 km/h when exerting a steady pull of 600 N. Calculate (a) the work done in 30 minutes and (b) the power required.

(a) Work done = force × distance moved.

Distance moved in 30 min, i.e. 0.5 h, at 90 km/h = 45 km

Hence, work done = 600 N × 45000 m = 27000 kJ or **27 MJ**

(b) Power required = $\dfrac{\text{work done}}{\text{time taken}}$ = $\dfrac{27 \times 10^{-6}\text{ J}}{30 \times 60\text{ s}}$ = 15000 W or **15 kW**

Example 4.16

To what height will a mass of weight 981 N be raised in 40 s by a machine using a power of 2 kW?

Work done = force × distance. Hence, work done = 981 N × height.

Power = (work done)/(time taken), from which,

work done = power × time taken = 2000 W × 40 s = 80000 J

Hence 80000 = 981 N × height, from which,

height = $\dfrac{80000\text{ J}}{981\text{ N}}$ = **81.55 m**

Potential and kinetic energy

Mechanical energy is concerned principally with two kinds of energy, potential energy and kinetic energy.

Potential energy is energy due to the position of the body. The force exerted on a mass of m kg is mg N (where g = 9.81 m/s^2, the acceleration due to gravity). When the mass is lifted vertically through a height h m above some datum level, the work done is given by: force × distance = $(mg)(h)$ J. This work done is stored as potential energy in the mass. Hence,

potential energy = mgh joules

(the potential energy at the datum level being taken as zero).

Kinetic energy is the energy due to the motion of a body. Suppose a force acts on an object of mass m originally at rest and accelerates it to a velocity v.

Then

$$\text{kinetic energy} = \tfrac{1}{2} mv^2 \text{ joules}$$

As stated earlier, energy may be converted from one form to another. The *principle of conservation of energy* states that the total amount of energy remains the same in such conversions, i.e. energy cannot be created or destroyed.

In mechanics, the potential energy possessed by a body is frequently converted into kinetic energy, and vice versa. When a mass is falling freely, its potential energy decreases as it loses height, and its kinetic energy increases as its velocity increases.

Example 4.17

A car of mass 800 kg is climbing an incline at 10° to the horizontal. Determine the increase in potential energy of the car as it moves a distance of 50 m up the incline.

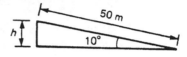

Figure 4.4

With reference to Figure 4.4, sin 10° = $h/50$, from which,

h = 50 sin 10° = 8.682 m

Hence increase in potential energy = mgh

$$= 800 \text{ k g} \times 9.81 \text{ m/s}^2 \times 8.682 \text{ m}$$

$$= 69140 \text{ J or } \textbf{68.14 kJ}$$

Example 4.18

At the instant of striking, a hammer of mass 30 kg has a velocity of 15 m/s. Determine the kinetic energy in the hammer.

$$\text{Kinetic energy} = \tfrac{1}{2} mv^2 = (30\text{kg})(15 \text{ m/s})^2$$

i.e. kinetic energy in hammer = **3375 J**

Electrical energy

Power P in an electrical circuit is given by the product of potential difference V and current I. The unit of power is the watt, W. Hence,

$$P = V \times I \text{ watts}$$

From Ohm's law (see later), $V = IR$. Substituting for V in the equation gives:

$$P = (IR) \times I = I^2R \quad \text{watts}$$

Similarly, $I = V/R$. Substituting for I in the equation gives:

$$P = V \times V/R = \frac{V^2}{R} \quad \text{watts}$$

There are thus *three* possible formulae that can be used for calculating power in electrical circuits.

The energy in an electrical circuit is the product of the power P and the time t. Hence,

Electrical energy = power × time

If the power is measured in watts and the time in seconds then the unit of energy is watt-seconds or *joules*.

If the power is measured in kilowatts and the time in hours then the unit of energy is kilowatt-hours, often called the 'unit of electricity'. An electricity meter in the home records the number of kilowatt-hours used and is thus an energy meter.

Example 4.19

A 12 V battery is connected across a load having a resistance of 40 Ω. Determine the current flowing in the load, the power consumed and the energy dissipated in 2 minutes.

Current, $I = \dfrac{V}{R} = \dfrac{12\,V}{40\,\Omega} = 0.3\,A$

Power consumed, $P = VI = (12)(0.3) = 3.6\,W$.

Energy dissipated = power × time = (3.6 W)(2 × 60 s) = **432 J**

(since 1 J = 1 W s).

Example 4.20

Electrical equipment in an office takes a current of 13 A from a 240 V supply. Estimate the cost per week of electricity if the equipment is used for 30 hours each week and 1 kWh of energy costs 7p.

Power = VI watts = 240 × 13 = 3120 W = 3.12 kW.

Energy used per week = power × time = (3.12 kW) × (30h) = 93.6 kW h.

Cost at 7p per kWh = 93.6 × 7 = 655.2 p.

Hence weekly cost of electricity = **£6.55**

Example 4.21

A source of 15 V supplies a current of 2 A for six minutes. How much energy is provided in this time?

Energy = power × time, and power = voltage × current.

Hence energy = VIt = 15 × 2 × (6 × 60) = 10800 W s = **10.8 kJ**

Example 4.22

A source of 15 V supplies a current of 2 A for six minutes. How much energy is provided in this time?

An electric heater consumes 3.6 MJ when connected to a 240 V supply for 40 minutes. Find the power rating of the heater and the current taken from the supply.

$$\text{Power} = \frac{\text{energy}}{\text{time}} = \frac{3 \times 10^6 \, \text{J}}{40 \times 60 \, \text{s}} \, (\text{or W}) = 1500 \, \text{W}$$

i.e. power rating of heater = **1.5 kW**

Power $P = VI$, thus

$$I = \frac{P}{V} = \frac{1500 \, \text{W}}{250 \, \text{V}} = 6 \, \text{A}$$

Hence the current taken from the supply is **6 A**.

Test your knowledge 4.7

1 Calculate the power dissipated when a current of 4 mA flows through a resistance of 5 kΩ.

2 A current of 5 A flows in the winding of an electric motor, the resistance of the winding being 100 Ω. Determine (a) the p.d. across the winding, and (b) the power dissipated by the coil.

3 Determine the power dissipated by the element of an electric fire of resistance 20 Ω when a current of 10 A flows through it. If the fire is on for 6 hours determine the energy used and the cost if 1 unit of electricity costs 7p.

4 A business uses two 3 kW fires for an average of 20 hours each per week, and six 150 W lights for 30 hours each per week. If the cost of electricity is 7p per unit, determine the weekly cost of electricity to the business.

Activity 4.4

Investigate the electrical energy that is supplied to your own home. Describe how this energy is used and how the supply is metered. Estimate the average total energy consumed in (a) a summer month and (b) a winter month. Present your findings in the form of a brief word processed report.

Activity 4.5

Invesigate the battery starting system used in a conventional motor car. Identify the electrical and mechanical components used in the system, explaining what each does and how they work together. Present your findings in the form of a brief class presentation using appropriate diagrams and visual aids.

Heat energy

Heat is another form of energy. It is also measured in joules. *Temperature* is the degree of hotness or coldness of a substance. Heat and temperature are thus *not* the same thing.

For example, twice the heat energy is needed to boil a full container of water than half a container – that is, different amounts of heat energy are needed to cause an equal rise in the temperature of different amounts of the same substance.

Temperature is measured either:

(i) on the *Celsius* ($^{\circ}$C) scale (formerly Centigrade), where the temperature at which ice melts, i.e. the freezing point of water, is taken as 0°C and the point at which water boils under normal atmospheric pressure is taken as 100°C, or

(ii) on the *thermodynamic scale*, in which the unit of temperature is the Kelvin (K). The Kelvin scale uses the same temperature interval as the Celsius scale but as its zero takes the 'absolute zero of temperature' which is at about −273°C. Hence,

$$\text{Kelvin temperature} = \text{degree Celsius} + 273$$

i.e.

$$K = (^{\circ}C) + 273$$

Thus, for example, 0°C = 273 K, 25°C = 298 K and 100° = 373 K.

Similarly,

$$\text{degree Celsius} = \text{Kelvin temperature} - 273$$

Example 4.23

Convert the following temperatures into the Kelvin scale: (a) 37$\circ$C (b) −28$\circ$C

From the above, Kelvin temperature = degree Celsius + 273

(a) 37$\circ$C corresponds to a Kelvin temperature of 37 + 273, i.e. **310 K**

(b) −28$\circ$C corresponds to a Kelvin temperature of −28 + 273, i.e. **245 K**

Example 4.24

Convert the following temperatures into the Celsius scale: (a) 365 K (b) 213 K

From above, K = ($\circ$C) + 273

Hence, degree Celsius = Kelvin temperature −273

(a) 365 K corresponds to 365 − 273, i.e. **92$\circ$C**

(b) 213 K corresponds to 213 − 273, i.e. **−60$\circ$C**

Specific heat capacity

The *specific heat capacity* of a substance is the quantity of heat energy required to raise the temperature of 1 kg of the substance by 1°C. The symbol used for specific heat capacity is c and the units are J/(kg °C) or J(kg K). (Note that these units may also be written as J kg^{-1} °C^{-1} or J kg^{-1} K^{-1}.)

Some typical values of specific heat capacity for the range of temperature 0°C to 100°C include:

Water	4190 J(kg °C)	Ice	2100 J/(kg °C)
Aluminium	950 J/(kg °C)	Copper	390 J/(kg °C)
Iron	500 J(kg °C)	Lead	130 J/(kg °C)

Hence to raise the temperature of 1 kg of iron by 1°C requires 500 J of energy, to raise the temperature of 5 kg of iron by 1°C requires (500×5) J of energy, and to raise the temperature of 5 kg of iron by 40°C requires $(500 \times 5 \times 40)$J of energy, i.e. 100 kJ.

In general, the quantity of heat energy, Q, required to raise a mass m kg of a substance with a specific heat capacity c J/(kg °C) from temperature t_1°C to t_2 °C is given by:

$$Q = mc(t_2 - t_1) \text{ joules}$$

Example 4.25

Calculate the quantity of heat required to raise the temperature of 5 kg of water from 0°C to 100°C. Assume the specific heat capacity of water is 4200 J/(kg °C)

Quantity of heat energy, $Q = mc(t_2 - t_1)$

$$= 5 \text{ kg} \times 4200 \text{ J/(kg °C)} \times (100 - 0)°C$$

$$= 5 \times 4200 \times 100$$

$$= 2100000 \text{ J or } 2100 \text{ kJ or } \textbf{2.1 MJ}$$

Example 4.26

A block of cast iron having a mass of 10 kg cools from a temperature of 150°C to 50°C. How much energy is lost by the cast iron? Assume the specific heat capacity of iron is 500 J/(kg °C).

Quantity of heat energy, $Q = mc(t_2 - t_1)$

$$= 10 \text{ kg x } 500 \text{ J/(kg °C) x } (50 - 150) °C$$

$$= 10 \text{ x } 500 \text{ x } (-100)$$

$$= -500000 \text{ J or } -500 \text{ kJ or } \textbf{-0.5 MJ}$$

(Note that the minus sign indicates that heat is given out or lost.)

1 (a) Convert –63°C into the Kelvin scale.
(b) Convert 225 K into the Celsius scale.

2 20.8 kJ of heat energy is required to raise the temperature of 2 kg of lead from 16°C to 96°C. Determine the specific heat capacity of lead.

3 5.7 MJ of heat energy are supplied to 30 kg of aluminium which is initially at a temperature of 20°C. If the specific heat capacity of aluminium is 950 J/(kg °C), determine its final temperature.

Example 4.25

Some lead having a specific heat capacity of 130 J/(kg °C) is heated from 27°C to its melting point at 327°C. If the quantity of heat required is 780 kJ determine the mass of the lead.

Quantity of heat, $Q = mc(t_2 - t_1)$,

hence 780×10^3 J $= m \times 130$ J/(kg °C) $\times (327 - 27)$ °C,

i.e. $780000 = m \times 130 \times 300$

from which,

$$\text{mass } m = \frac{780000}{130 \times 300} \text{ kg} = \textbf{20 kg}$$

Change of state

A material may exist in any one of three states – solid, liquid or gas. If heat is supplied at a constant rate to some ice initially at, say, –30°C, its temperature rises as shown in Figure 4.5. Initially the temperature increases from –30°C to 0°C as shown by the line AB. It then remains constant at 0°C for the time BC required for the ice to melt into water.

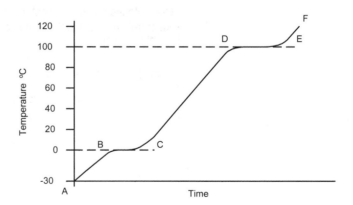

Figure 4.5

When melting commences the energy gained by continual heating is offset by the energy required for the change of state and the temperature remains constant even though heating is continued. When the ice is completely melted to water, continual heating raises the temperature to 100°C, as shown by CD in Figure 4.5. The water then begins to boil and the temperature again remains constant at 100°C, shown as DE, until all the water has vaporized.
Continual heating raises the temperature of the steam as shown by EF in the region where the steam is termed superheated.

Changes of state from solid to liquid or liquid to gas occur without change of temperature and such changes are reversible processes.

When heat energy flows to or from a substance and causes a change of temperature, such as between A and B, between C and D and between E and F in Figure 4.5, it is called *sensible heat* (since it can be 'sensed' by a thermometer).

Heat energy that flows to or from a substance while the temperature remains constant, such as between B and C and between D and E in Figure 4.5, is called *latent heat* (latent means concealed or hidden).

Latent heat

The *specific latent heat of fusion* is the heat required to change 1 kg of a substance from the solid state to the liquid state (or vice versa) at constant temperature.

The *specific latent heat of vaporization* is the heat required to change 1 kg of a substance from a liquid to a gaseous state (or vice versa) at constant temperature. The units of the specific latent heats of fusion and vaporization are J/kg, or more often kJ/kg, and some typical values are shown in Table 4.1. The quantity of heat Q supplied or given out during a change of state is given by:

$$Q = mL$$

where m is the mass in kilograms and L is the specific latent heat.

Thus, for example, the heat required to convert 10 kg of ice at 0°C to water at 0°C is given by 10 kg × 335 kJ/kg, i.e. 3350 kJ or 3.35 MJ.

Besides changing temperature, the effects of supplying heat to a material can involve changes in dimensions, as well as in colour, state and electrical resistance. Most substances expand when heated and contract when cooled, and there are many practical applications and design implications of thermal movement.

Table 4.1

	Latent heat of fusion (kJ/kg)	*Melting point (°C)*
Mercury	11.8	−39
Lead	22	327
Silver	100	957
Ice	335	0
Aluminium	387	660
	Latent heat of vaporization (kJ/kg)	*Boiling point (°C)*
Oxygen	214	−183
Mercury	286	357
Ethyl alcohol	857	79
Water	2257	100

1 Determine the heat energy required to change 8 kg of water at 100°C to superheated steam at 100°C. Assume the specific latent heat of vaporization of water is 2260 kJ/kg.

2 Calculate the heat energy required to convert completely 10 kg of water at 50°C into steam at 100°C, given that the specific heat capacity of water is 4200 J/(kg°C) and the specific latent heat of vaporization of water is 2260 kJ/kg.

Example 4.28

How much heat is needed to melt completely 12 kg of ice at 0°C? Assume the latent heat of fusion of ice is 335 kJ/kg.

Quantity of heat required, $Q = mL$

$$= 12 \text{ kg} \times 335 \text{ kJ/kg}$$

$$= 4020 \text{ kJ or } \mathbf{4.02 \text{ MJ}}$$

Example 4.29

Calculate the heat required to convert 5 kg of water at 100°C to superheated steam at 100°C. Assume the latent heat of vaporization of water is 2260 kJ/kg.

Quantity of heat required, $Q = mL$

$$= 5 \text{ kg} \times 2260 \text{ kJ/kg}$$

$$= 11300 \text{ kJ or } \mathbf{11.3 \text{ MJ}}$$

Example 4.30

Determine the heat energy needed to convert 5 kg of ice initially at −20°C completely to water at 0°C. Assume the specific heat capacity of ice is 2100 J/(kg °C) and the specific latent heat of fusion of ice is 335 kJ/kg.

Quantity of heat energy needed, Q = sensible heat + latent heat.

The quantity of heat needed to raise the temperature of ice from −20°C to 0°C, i.e. sensible heat, is given by

$$Q_1 = mc(t_2 - t_1)$$

$$= 5 \text{ kg} \times 2100 \text{ J/(kg °C)} \times (0 - -20)°C$$

$$= (5 \times 2100 \times (+20)) \text{J}$$

$$= 210 \text{ kJ}$$

The quantity of heat needed to melt 5 kg of ice at 0°C, i.e. the latent heat,

$$Q_2 = mL = 5 \text{ kg} \times 335 \text{ kJ/kg} = 1675 \text{ kJ}.$$

Total heat energy needed,

$$Q = Q_1 + Q_2 = 210 + 1675 = \mathbf{1885 \text{ kJ}}$$

Electrical applications

All *atoms* consist of *protons*, *neutrons* and *electrons*. The protons, which have positive electrical charges, and the neutrons, which have no electrical charge, are contained within the *nucleus*. Removed from the nucleus are minute negatively charged particles called electrons. Atoms of different materials differ from one another by having different numbers of protons, neutrons and electrons. An equal number of protons and electrons exist within an atom and it is said to be electrically balanced, as the positive and negative charges cancel each other out. When there are more than two electrons in an atom the electrons are arranged into *shells* at various distances from the nucleus.

All atoms are bound together by powerful forces of attraction existing between the nucleus and its electrons. Electrons in the outer shell of an atom, however, are attracted to their nucleus less powerfully than are electrons whose shells are nearer the nucleus.

It is possible for an atom to lose an electron; the atom, which is now called an *ion*, is not now electrically balanced, but is positively charged and is thus able to attract an electron to itself from another atom. Electrons that move from one atom to another are called free electrons and such random motion can continue indefinitely. However, if an electric pressure or *voltage* is applied across any material there is a tendency for electrons to move in a particular direction.

This movement of free electrons, known as *drift*, constitutes an electric current flow. Thus *current is the rate of movement of charge.*

Conductors are materials that have electrons that are loosely connected to the nucleus and can easily move through the material from one atom to another. *Insulators* are materials whose electrons are held firmly to their nucleus.

The unit used to measure the *quantity of electrical charge Q* is called the *coulomb* (C) (where 1 coulomb = 6.24×10^{18} electrons). If the drift of electrons in a conductor takes place at the rate of one coulomb per second the resulting current is said to be a current of one ampere. Thus,

1 ampere = 1 coulomb per second

or

1 A = 1 C/s

Hence

1 coulomb = 1 ampere second   or   1 C = 1 A s

Generally, if I is the current in amperes and t the time in seconds during which the current flows, then $I \times t$ represents the quantity of electrical charge in coulombs, i.e. quantity of electrical charge transferred.

$Q = I \times t$ *coulombs*

Example 4.31

If a current of 10 A flows for four minutes, find the quantity of electricity transferred.

Quantity of electricity, $Q = It$ coulombs. $I = 10$ A, $t = 4 \times 60 = 240$ s.

Hence $Q = 10 \times 240 =$ **2400 C**

Potential difference and resistance

For a continuous current to flow between two points in a circuit a *potential difference* (p.d.) or *voltage* (V) is required between them; a complete conducting path is necessary to and from the source of electrical energy. The unit of p.d. is the volt (V). Figure 4.6 shows a cell connected across a filament lamp. Current flow, by convention, is considered as flowing from the positive terminal of the cell around the circuit to the negative terminal.

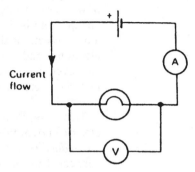

Figure 4.6

The flow of electric current is subject to friction. This friction, or opposition, is called *resistance* (R) and is the property of a conductor that limits current. The unit of resistance is the *ohm* (Ω). 1 ohm is defined as the resistance that will have a current of 1 ampere flowing through it when 1 volt is connected across it, i.e.

$$\text{resistance, } R = \frac{\text{potential difference, } V}{\text{current, } I}$$

Potential difference and resistance

Ohm's law states that the current I flowing in a circuit is directly proportional to the applied voltage V and inversely proportional to the resistance R, provided the temperature remains constant. Thus,

$$I = \frac{V}{R} \quad \text{or} \quad V = IR \quad \text{or} \quad R = \frac{V}{I}$$

Test your knowledge 4.10

1 What current must flow if 0.24 coulombs is to be transferred in 15 ms?

2 A p.d. of 50 V is applied across a heating element. If the resistance of the element is 12.5 Ω, find the current flowing through it.

Example 4.32

The current flowing through a resistor is 0.8 A when a p.d. of 20 V is applied. Determine the value of the resistance.

From Ohm's law,

$$\text{resistance } R = \frac{V}{I} = \frac{20}{0.8} = \textbf{25} \, \boldsymbol{\Omega}$$

Example 4.33

Determine the p.d. which must be applied to a 2 kΩ resistor in order that a current of 10 mA may flow.

Resistance R = 2 kΩ = 2 x 10³ = 2000 Ω

Current I = 10 mA = 10 x 10⁻³ A = 0.01 A

From Ohm's law, potential difference, $V = IR = (0.01)(2000) =$ **20 V**

Test your knowledge 4.11

1 A 100 V battery is connected across a resistor and causes a current of 5 mA to flow. Determine the resistance of the resistor. If the voltage is now reduced to 25 V, what will be the new value of the current flowing?

2 What is the resistance of a coil which draws a current of (a) 50 mA and (b) 200 μA from a 120 V supply?

Example 4.34

A coil has a current of 50 mA flowing through it when the applied voltage is 12 V. What is the resistance of the coil?

$$\text{Resistance, } R = \frac{V}{I} = \frac{12}{50 \times 10^{-3}} = \textbf{240 Ω}$$

Example 4.35

A 100 W electric light bulb is connected to a 250 V supply. Determine (a) the current flowing in the bulb, and (b) the resistance of the bulb.

From earlier work,

power $P = V \times I$,   from which, current $I = \dfrac{P}{V}$

(a) Current $I = \dfrac{100}{250} =$ **0.4 A**

(b) Resistance $R = \dfrac{250}{0.4} =$ **625 Ω**

Example 4.36

An electric kettle has a resistance of 30 Ω. What current will flow when it is connected to a 240 V supply? Find also the power rating of the kettle.

$$\text{Current, } I = \frac{V}{R} = \frac{240}{30} = \textbf{8 A}$$

Power rating of kettle, $P = VI = 240 \times 8 = 1920$ W = **1.92 kW**

Example 4.37

The current/voltage relationship for two resistors A and B is as shown in Figure 4.7. Determine the value of the resistance of each resistor.

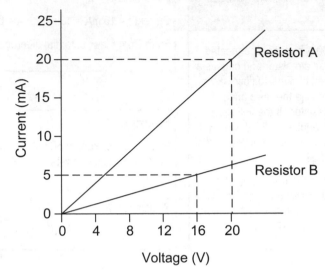

Figure 4.7

For resistor A,

$$R = \frac{V}{I} = \frac{20 \text{ V}}{20 \text{ mA}} = \frac{20 \text{ V}}{0.02 \text{ A}} = 1000 \text{ }\Omega \text{ or } \mathbf{1 \text{ k}\Omega}$$

For resistor B,

$$R = \frac{V}{I} = \frac{16 \text{ V}}{5 \text{ mA}} = \frac{16 \text{ V}}{0.005 \text{ A}} = 3200 \text{ }\Omega \text{ or } \mathbf{3.02 \text{ k}\Omega}$$

Example 4.38

The hot resistance of a 240 V filament lamp is 960 Ω. Find the current taken by the lamp and its power rating.

From Ohm's law,

$$\text{current } I = \frac{V}{R} = \frac{240 \text{ V}}{960 \text{ }\Omega} = \mathbf{0.25 \text{ A}}$$

Power rating $P = VI = 240 \text{ V} \times 0.25 \text{ A} = \mathbf{60 \text{ W}}$

Resistance and resistivity

The resistance of an electrical conductor depends on four factors, these being: (a) the length of the conductor, (b) the cross-sectional area of the conductor, (c) the type of material and (d) the temperature of the material.

Resistance, R, is directly proportional to length, l, of a conductor, i.e. $R \propto l$. Thus, for example, if the length of a piece of wire is doubled, then the resistance is doubled. Resistance, R, is inversely proportional to cross-sectional area, a, of a conductor, i.e. $R \propto l/a$. Thus, for example, if the cross-sectional area of a piece of wire is doubled then the resistance is halved.

Since $R \propto l$ and $R \propto 1/a$ then $R \propto l/a$. By inserting a constant of proportionality into this relationship the type of material used may be taken into account. The constant of proportionality is known as the *resistivity* of the material and is given the symbol ρ (rho). Thus

resistance $R = \dfrac{\rho\, l}{a}$ ohms

ρ is measured in ohm metres (Ω m). The value of the resistivity is the resistance of a unit cube of the material measured between opposite faces of the cube.

Resistivity varies with temperature and some typical values of resistivities measured at about room temperature are given in the table below. Note that good conductors of electricity have a low value of resistivity whereas good insulators have a high value of resistivity.

Table 4.2 *Resistivity of some common materials*

Material	Resistivity (Ω m)
Copper	1.7×10^{-8}
Aluminium	2.6×10^{-8}
Carbon (graphite)	10×10^{-8}
Glass	1×10^{10}
Mica	1×10^{13}

Example 4.39

Calculate the resistance of a 2 km length of aluminium overhead power cable if the cross-sectional area of the cable is 100 mm². Take the resistivity of aluminium to be 0.03 x 10⁻⁶ Ω m.

Length l = 2 km = 2000 m; area, a = 100 mm² = 100 x 10⁻⁶ m²; resistivity ρ = 0.03 x 10⁻⁶ Ω m

Resistance $R = \dfrac{\rho l}{a} = \dfrac{(0.03 \times 10^{-6}\ \Omega m)(2000\ m)}{(100 \times 10^{-6}\ m^2)} = \mathbf{0.6\ \Omega}$

Example 4.40

The resistance of a 5 m length of wire is 600 Ω. Determine (a) the resistance of an 8 m length of the same wire, and (b) the length of the same wire when the resistance is 420 Ω.

(a) Resistance, R, is directly proportional to length, l, i.e. $R \propto l$. Hence,

600 Ω $\propto$ 5 m or 600 = $(k)(5)$, where k is the coefficient of proportionality.

Hence,

$$k = \frac{600}{5} = 120$$

When the length l is 8 m, then

resistance $R = kl = (120)(8) = \mathbf{960\ \Omega}$

(b) When the resistance is 420 Ω, 420 = kl, from which

$$\text{length } l = \frac{420}{k} = \frac{420}{120} = \mathbf{3.5\ m}$$

Series circuits

Figure 4.8 shows three resistors, R_1, R_2 and R_3 connected end to end, i.e. in *series*, with a battery source of V volts. Since the circuit is closed a current I will flow and the p.d. across each resistor may be determined from the voltmeter readings V_1, V_2 and V_3.

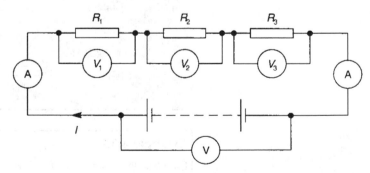

Figure 4.8

In a series circuit:

(a) the current I is the same in all parts of the circuit and hence the same reading is found on each of the ammeters shown, and

(b) the sum of the voltages V_1, V_2 and V_3 is equal to the total applied voltage, V, i.e.

$$V = V_1 + V_2 + V_3$$

From Ohm's law:

$$V_1 = IR_1, V_2 = IR_2, V_3 = IR_3 \text{ and } V = IR$$

where R is the total circuit resistance.

Since $V = V_1 + V_2 + V_3$, then $IR = IR_1 + IR_2 + IR_3$. Dividing throughout by I gives

$$R = R_1 + R_2 + R_3$$

Thus for a series circuit, the total resistance is obtained by adding together the values of the separate resistances.

Example 4.41

For the circuit shown in Figure 4.9, determine (a) the battery voltage V, (b) the total resistance of the circuit, and (c) the values of resistance of resistors R_1, R_2 and R_3, given that the p.d.s across R_1, R_2 and R_3 are 5 V, 2 V and 6 V, respectively.

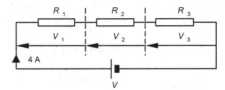

Figure 4.9

(a) Battery voltage $V = V_1 + V_2 + V_3 = 5 + 2 + 6 = $ **13 V**

(b) Total circuit resistance $R = \dfrac{V}{I} = \dfrac{13 \text{ V}}{4 \text{ A}} = $ **3.25 Ω**

(c) Resistance $R_1 = \dfrac{V_1}{I} = \dfrac{5 \text{ V}}{4 \text{ A}} = $ **1.25 Ω**

Resistance $R_2 = \dfrac{V_2}{I} = \dfrac{2 \text{ V}}{4 \text{ A}} = $ **0.5 Ω**

Resistance $R_3 = \dfrac{V_3}{I} = \dfrac{6 \text{ V}}{4 \text{ A}} = $ **1.5 Ω**

(Check: $R_1 + R_2 + R_3 = 1.25 + 0.5 + 1.5 = 3.25 \text{ Ω} = R$)

Test your knowledge 4.13

A 12 V battery is connected in a circuit having three series-connected resistors having resistance of 4 Ω, 9 Ω and 11 Ω. Determine the current flowing through, and the p.d. across, the 9 Ω resistor. Find also the power dissipated in the 11 Ω resistor.

Example 4.42

Resistors of 10 Ω, 15Ω and 25 Ω are connected in series across a supply of 25 V. Determine the total resistance of the circuit and the current supplied.

Total resistance $R = R_1 + R_2 + R_3 = 10 + 15 + 25 = 50 \text{ Ω}$

Current flowing $I = \dfrac{V}{R} = \dfrac{25}{50} = 0.5 \text{ A}$

Parallel circuits

Figure 4.10 shows three resistors, R_1, R_2 and R_3 connected across each other, i.e. in parallel, across a battery source of V volts.

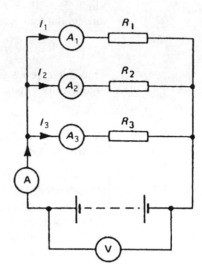

Figure 4.10

In a parallel circuit:

(a) the sum of the currents I_1, I_2 and I_3 is equal to the total circuit current, I, i.e.

$$I = I_1 + I_2 + I_3$$

(b) the source p.d., V volts, is the same across each of the resistors.

From Ohm's law:

$$I_1 = V/R_1 , \quad I_2 = V/R_2 , I_3 = V/R_3 \quad \text{and} \quad I = V/R$$

where R is the total circuit resistance. Since

$$I = I_1 + I_2 + I_3$$

then

$$V/R = V/R_1 + V/R_2 + V/R_3$$

Dividing throughout by V gives

$$1/R = 1/R_1 + 1/R_2 + 1/R_3$$

This equation must be used when finding the total resistance R of a parallel circuit. For the special case of *two resistors in parallel*

$$1/R = 1/R_1 + 1/R_2 = \frac{R_1 + R_2}{R_1 R_2}$$

Thus $R = \dfrac{R_1 R_2}{R_1 + R_2} = \dfrac{product}{sum}$

Example 4.43

For the circuit shown in Figure 4.11, determine (a) the reading on the ammeter, and (b) the value of resistor R_2

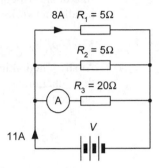

Figure 4.11

The p.d. across R_1 is the same as the supply voltage V.

Hence supply voltage, $V = 8 \times 5 = 40$ V.

(a) Reading on ammeter

$$I = \frac{V}{R_3} = \frac{40\ V}{20\ \Omega} = \textbf{2 A}$$

(b) Current flowing through $R_2 = 11 - 8 - 2 = 1$ A, hence

$$R_2 = \frac{V}{I_2} = \frac{40\ V}{1\ A} = \textbf{40 } \boldsymbol{\Omega}$$

Test your knowledge 4.14

1 Two resistors, of resistance 3 Ω and 6 Ω, are connected in parallel across a battery having a voltage of 12 V. Determine (a) the total circuit resistance and (b) the current flowing in the 3 Ω resistor.

2 Given four 1 Ω resistors, state how they must be connected to give an overall resistance of (a) 0.25 Ω, (b) 1 Ω, (c) 1.33 Ω, (d) 2.5 Ω, all four resistors being connected in each case.

3 Resistances of 10 Ω, 20 Ω and 30 Ω are connected (a) in series and (b) in parallel to a 240 V supply. Calculate the supply current in each case.

Example 4.44

Find the equivalent resistance for the circuit shown in Figure 4.12.

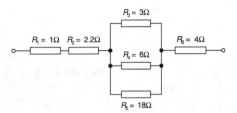

Figure 4.12

R_3, R_4 and R_5 are connected in parallel and their equivalent resistance R is given by:

$$1/R = 1/R_1 + 1/R_2 + 1/R_3 = 1/3 + 1/6 + 1/18 = \frac{6 + 3 + 1}{18} = \frac{10}{18}$$

Hence $R = (18/10) = 1.8\ \Omega$. The circuit is now equivalent to four resistors in series and the equivalent circuit resistance $= 1 + 2.2 + 1.8 + 4 = 9\ \Omega$.

Kirchhoff's laws

Kirchhoff (a German physicist) developed two laws that help us to find unknown currents and voltages in more complicated d.c. series/parallel networks.

Kirchhoff's current law

At any junction in an electric circuit the total current flowing towards that junction is equal to the total current flowing away from the junction, i.e. $\Sigma I = 0$.

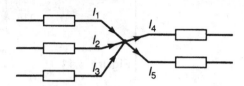

Figure 4.13

Thus referring to Figure 4.13

$$I_1 + I_2 + I_3 = I_4 + I_5$$

or

$$I_1 + I_2 + I_3 - I_4 - I_5 = 0$$

Kirchhoff's voltage law

In any closed loop in a network, the algebraic sum of the voltage drops (i.e. products of current and resistance) taken around the loop is equal to the resultant e.m.f. acting in that loop.

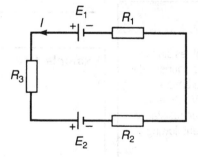

Figure 4.14

Thus referring to Figure 4.14:

$$E_1 - E_2 = IR_2 + IR_2 + IR_3$$

(Note that if current flows away from the positive terminal of a source, that source is considered by convention to be positive. Thus moving anticlockwise around the loop of Figure 4.14, E_1 is positive and E_2 is negative.)

Example 4.45

Determine the value of the unknown currents marked in Figure 4.15.

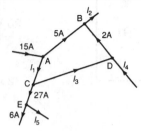

Figure 4.15

Applying Kirchhoff's current law to each junction in turn gives:

For junction A: $15 = 5 + I_1$

Hence $I_1 = \mathbf{10\ A}$

For junction B: $5 + 2 = I_2$

Hence $I_2 = \mathbf{7\ A}$

For junction C: $I_1 = 27 + I_3$ i.e. $10 = 27 + I_3$

Hence $I_3 = 10 - 27 = \mathbf{-17\ A}$

(i.e. in the opposite direction to that shown in Figure 4.15).

For junction D: $I_3 + I_4 = 2$ i.e. $-17 + I_4 = 2$

Hence $I_4 = 17 + 2 = \mathbf{19\ A}$

For junction E: $27 = 6 + I_5$

Hence $I_5 = 27 - 6 = \mathbf{21\ A}$

Example 4.46

Determine the value of e.m.f. E in Figure 4.16.

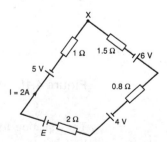

Figure 4.16

Applying Kirchhoff's voltage law and moving clockwise around the loop of Figure 4.16 starting at point X gives:

$6 + 4 + E - 5 = I(1.5) + I(0.8) + I(2) + I(1)$

$5 + E = I(5.3) = 2(5.3)$ since current I is 2 A.

Hence $5 + E = 10.6$ and e.m.f. $E = 10.6 - 5 = \mathbf{5.6\ V}$

Test your knowledge 4.15

Use Kirchhoff's laws to determine the current flowing in each branch of the network shown in Figure 4.17.

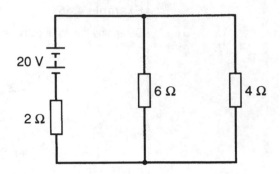

Figure 4.17

Alternating voltage and current

Unlike direct currents that have steady values and always flow in the same direction, alternating currents flow alternately one way and then the other. The alternating potential difference (voltage) produced by an alternating current is thus partly positive and partly negative. An understanding of alternating currents and voltages is important in a number of applications including a.c. power distribution, amplifiers and filters.

A graph showing the variation of voltage or current present in a circuit is known as a *waveform*. The most common waveform is the sine wave, see Figure 4.18.

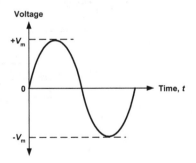

Figure 4.18 *A sinusoidal voltage*

The equation for the sinusoidal voltage shown in Figure 4.18, at a time, *t*, is:

$$v = V_m \sin(\omega t)$$

where v is the instantaneous voltage, V_m is the *maximum value* of voltage (also known as the *amplitude* or *peak value* of the voltage), and ω is the *angular velocity* (in radians per second).

The frequency of a waveform is the number of cycles of the waveform which occur in unit time. Frequency is expressed in hertz (Hz) and a frequency of 1 Hz is equivalent to one cycle per second.

Hence, if a voltage has a frequency of 50 Hz, fifty cycles will occur in every second.

Since there are 2 π radians in one complete revolution or cycle, a frequency of one cycle per second must be the same as 2 π radians per second. Hence a frequency *f* is equivalent to:

$$f = \omega / (2\,\pi) \quad \text{Hz}$$

Alternatively, the angular velocity, ω, is given by:

$$\omega = 2\,\pi\,f \quad \text{radians per second}$$

We can thus express the instantaneous voltage in another way:

$$v = V_{max} \sin(2\,\pi\,f\,t)$$

Example 4.47

A sine wave voltage has a maximum value of 100 V and a frequency of 50 Hz. Determine the instantaneous voltage present (a) 2.5 ms and (b) 15 ms from the start of the cycle.

We can determine the voltage at any instant of time using:

$$v = V_m \sin(2\,\pi\,f\,t)$$

where V_m = 100 V and *f* = 50 Hz

In (a), *t* = 2.5ms hence:

$$v = 100 \sin(2\,\pi \times 50 \times 0.0025) = 100 \sin(0.785) = 100 \times 0.707 = \mathbf{70.7\ V}$$

In (b), *t* = 15ms hence:

$$v = 100 \sin(2\,\pi \times 50 \times 0.015) = 100 \sin(4.71) = 100 \times -1 = \mathbf{-100\ V}$$

Periodic time

The periodic time of a waveform is the time taken for one complete cycle of the wave. The relationship between periodic time and frequency is thus:

$$t = 1/f \quad \text{or} \quad f = 1/t$$

where *t* is the periodic time (in seconds) and *f* is the frequency (in Hz).

From this relationship it is worth noting the periodic time of a waveform *decreases* as its frequency *increases*, and vice versa.

Example 4.48

A waveform has a frequency of 200 Hz. What is the periodic time of the waveform?

$$t = 1/f = 1/200 = 0.005\ \text{s} = \mathbf{5\ ms}$$

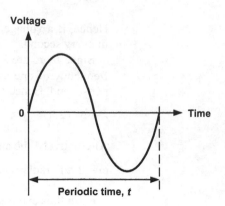

Figure 4.19 *Periodic time of a sinusoidal waveform*

Example 4.49

A waveform has a periodic time of 2.5ms. What is its frequency?

$f = 1 / t = 1 / (2.5 \times 10^{-3})$ = **400 Hz**

Average, r.m.s. and peak–peak values

The *average* value of an alternating current that swings symmetrically above and below zero will obviously be zero when measured over a long period of time. Hence average values of currents and voltages are invariably taken over one complete half-cycle (either positive or negative) rather than over one complete full-cycle (which would result in an average value of zero).

The *peak* value (or *amplitude*) of a waveform is a measure of the extent of its voltage or current excursion from the resting value

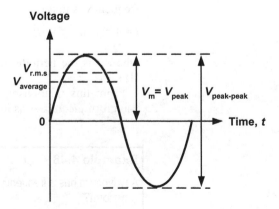

Figure 4.20 *Average, r.m.s. and peak–peak values*

(usually zero). The *peak–peak* value for a wave that is symmetrical about its resting value is twice its peak value.

The *r.m.s.* (or *effective*) value of an alternating voltage or current is the value which would produce the same heat energy in a resistor as a direct voltage or current of the same magnitude. Since the r.m.s. value of a waveform is very much dependent upon its shape, values are only meaningful when dealing with a waveform of known shape. Where the shape of a waveform is not specified, r.m.s. values are normally assumed to refer to sinusoidal conditions.

The following formulae apply to a sine wave:

$$V_{average} \; = \; \frac{2}{\pi} \times V_m = 0.636 \times V_m$$

$$V_{peak-peak} = 2 \times V_m$$

$$V_{r.m.s.} \; = \; \frac{1}{\sqrt{2}} \times V_m = 0.707 \times V_m$$

Similar relationships apply to the corresponding alternating currents, thus:

$$I_{average} = 0.636 \times I_m = 0.636 \times I_m$$

$$I_{peak-peak} = 2 \times I_m$$

$$I_{r.m.s.} = \times I_m = 0.707 \times I_m$$

Example 4.50

A sinusoidal voltage has an r.m.s. value of 220 V. What is the peak value of the voltage?

Since $V_{r.m.s.} = 0.707 \, V_m$, $V_m = V_{r.m.s.} / 0.707 = 1.414 \times V_{r.m.s.}$

Thus $V_m = 1.414 \times 220 = \mathbf{311 \; V}$

Test your knowledge 4.17

1 Determine the r.m.s. value of a sinusoidal current that has a maximum value of 50 mA.

2 Determine the peak–peak value of a sinusoidal voltage that has an r.m.s. value of 110 V.

3 Determine the r.m.s. value of a sinusoidal current that has an average value of 2 A.

Example 4.51

A sinusoidal alternating current has a peak-peak value of 4 mA. What is its r.m.s. value?

First we must convert the peak–peak current into peak current:

Since $I_{peak-peak} = 2 \times I_m$, $I_m = 0.5 \times I_{peak-peak}$

Thus $I_{peak} = 0.5 \times 4 = 2$ mA

Now we can convert the peak current into r.m.s. current using:

$I_{r.m.s.} = 0.707 \times I_{peak}$

Thus $I_{r.m.s.} = 0.707 \times 2 = \mathbf{1.414 \; mA}$

Alternating current in a resistor

Ohm's Law is obeyed in an a.c. circuit just as it is in a d.c. circuit. Thus, when a sinusoidal voltage V is applied to a resistor R (as shown in Figure 4.21) the current flowing in the resistor will be given by:

$$I = \frac{V}{R}$$

This relationship must also hold true for the instantaneous values of current i and voltage v thus:

$$i = \frac{v}{R}$$

and since $v = V_m \sin(\omega t)$

$$i = \frac{V_m \sin(\omega t)}{R}$$

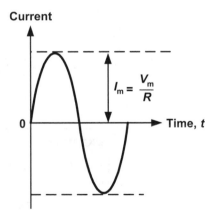

Figure 4.21 *Alternating current in a resistor*

Test your knowledge 4.18

1 Determine the power dissipated in a resistor of 50 Ω when an a.c. voltage having a peak–peak value of 100 V is applied to it.

2 A resistor has a power rating of 4 W. What is the maximum value of r.m.s. current that can be safely applied to it?

The power dissipated in the resistor will be given by the product of the r.m.s. current and voltage. Thus:

$$P = V_{r.m.s.} \times I_{r.m.s.} \text{ watts}$$

Once again recall that, from Ohm's law, $V_{r.m.s.} = I_{r.m.s.}R$ thus we can obtain two further relationships for the power dissipated:

$$P = I_{r.m.s.}{}^2 R \quad \text{watts}$$

and

$$P = \frac{V_{r.m.s.}{}^2}{R} \text{ watts.}$$

Mechanical applications

Velocity

You will recall that speed can be defined as the rate of covering distance. Thus,

$$\text{speed} = \frac{\text{distance travelled}}{\text{time taken}}$$

The *velocity* of an object is the speed of the object *in a specified direction*. Thus,

$$\text{velocity} = \frac{\text{distance travelled in a } \textit{specified direction}}{\text{time taken}}$$

A graph of velocity (vertical axis) against time (horizontal axis) is called a *velocity/time graph*. The graph shown in Figure 4.22 represents an aircraft flying for 3 hours at a constant speed of 600 kilometres per hour in a specified direction. The shaded area represents velocity (vertically) multiplied by time (horizontally), and has units of

$$\frac{\text{kilometres}}{\text{hours}} \times \text{hours} = \text{kilometres}$$

i.e. the distance travelled in a specified direction. In this case,

$$\text{distance} = 600 \text{ km/h} \times 3 \text{ h} = 1800 \text{ km}$$

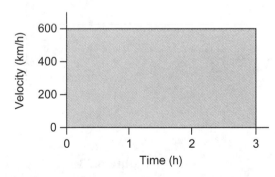

Figure 4.22

Test your knowledge 4.19

A car travels from town A to town B, a distance of 40 kilometres at an average speed of 55 kilometres per hour. It then travels from town B to town C, a distance of 25 kilometres in 35 minutes. Finally, it travels from town C to town D at an average speed of 60 kilometres per hour in 45 minutes. Determine:

(a) the time taken to travel from A to B,

(b) the average speed of the car from B to C,

(c) the distance from C to D, and

(d) the average speed of the whole journey from A to D.

Another method of determining the distance travelled is from:

$$\text{distance travelled} = \text{average velocity} \times \text{time}$$

Thus if a plane travels due south at 600 kilometres per hour for 20 minutes, the distance covered is

$$\frac{600 \text{ km}}{1 \text{ h}} \times \frac{20}{60} \text{ h} = 200 \text{ km}$$

Acceleration

Acceleration is defined as the rate of change of velocity with time. The average acceleration a is given by:

$$a = \frac{\text{change in velocity}}{\text{time taken}}$$

The usual units are metres per second squared (m/s^2 or ms^{-2}). If u is the initial velocity of an object in metres per second, v is the final velocity in metres per second and t is the time in seconds elapsing between the velocities of u and v, then

$$\text{average acceleration, } a = \frac{v - u}{t} \text{ m/s}^2$$

From the velocity/time graph shown in Figure 4.23, the slope of line OA is given by AX/OX. AX is the change in velocity from an initial velocity u of zero to a final velocity, v, of 4 metres per second. OX is the time taken for this change in velocity, thus

$$\frac{\text{change in velocity}}{\text{time taken}} = \frac{\text{AX}}{\text{OX}}$$

$$= \text{acceleration in the first two seconds}$$

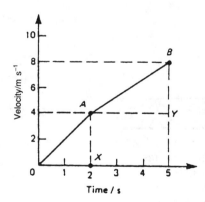

Figure 4.23

From the graph:

$$\frac{\text{AX}}{\text{OX}} = \frac{4 \text{ m/s}}{2 \text{ s}} = 2 \text{ m/s}^2$$

i.e. the acceleration is 2 m/s^2. Similarly, the slope of line AB in Figure 4.23 is given by BY/AY, i.e. the acceleration between 2 s and 5 s is

$$\frac{8 - 4}{5 - 2} = \frac{4}{3} = 1.33 \text{ m/s}^2$$

In general, the slope of a line on a velocity/time graph gives the acceleration. It is important to note that the words 'velocity' and 'speed' are commonly interchanged in everyday language. Acceleration is a vector quantity and is correctly defined as the rate of change of velocity with respect to time. However, acceleration is also the rate of change of speed with respect to time in a certain specified direction.

Free-fall

If a dense object such as a stone is dropped from a height, called *free-fall*, it has a constant acceleration of approximately 9.81 m/s^2. In a vacuum, all objects have this same constant acceleration, vertically downwards, that is, a feather has the same acceleration as a stone. However, if free-fall takes place in air, dense objects have the constant acceleration of 9.81 m/s^2 over short distances, but objects that have a low density, such as feathers, have little or no acceleration.

Equations of motion

For bodies moving with a constant acceleration, the average acceleration is the constant value of the acceleration, and since from earlier:

average acceleration, $a = $ m/s^2

then

$$a \times t = v - u \quad \text{or} \quad v = u + at$$

where u is the initial velocity in m/s, v is the final velocity in m/s, a is the constant acceleration in m/s^2, and t is the time in s.

When the acceleration a has a negative value, it is called *deceleration* or *retardation*. The equation $v = u + at$ is called an *equation of motion*.

Test your knowledge 4.20

1 A ship changes velocity from 15 km/h to 20 km/h in 25 min. Determine the average acceleration in m/s^2 of the ship during this time.

2 Determine how long it takes an object, which is free-falling, to change its speed from 100 km/h to 150 km/h, assuming all other forces, except that due to gravity, are neglected.

3 A car travelling at 50 km/h applies its brakes for 6 s and decelerates uniformly at 0.5 m/s. Determine its velocity in km/h after the 6 s braking period.

Example 4.52

A crate is dropped from an aircraft. Determine (a) its velocity after 2 s and (b) the increase in velocity during the third second, in the absence of all forces except that due to gravity.

The stone is free-falling and thus has an acceleration, a, of approximately 9.81 m/s^2 (taking downward motion as positive). From the equation of motion, final velocity, $v = u + at$

(a) The initial downward velocity of the stone, u, is zero. The acceleration, a, is 9.81 m/s^2 downwards and the time during which the stone is accelerating is 2 s. Hence, final velocity, $v = 0 + 9.81 \times 2 = 19.62$ m/s, i.e. the velocity of the stone after 2 s is approximately **19.62 m/s.**

(b) From part (a), the velocity after two seconds, u, is 19.62 m/s. The velocity after 3 s, applying $v = u + at$, is $v = 19.62 + 9.81 \times 3 = 49.05$ m/s. Thus, the change in velocity during the third second is (49.05 - 19.62) m/s or **29.43 m/s.**

Force, mass and acceleration

When an object is pushed or pulled, a force is applied to the object. This force is measured in newtons (N). The effects of pushing or pulling an object are:

(i) to cause a change in the motion of the object, and

(ii) to cause a change in the shape of the object.

If a change in the motion of the object, that is, its velocity changes from u to v, then the object accelerates. Thus, it follows that acceleration results from a force being applied to an object. If a force is applied to an object and it does not move, then the object changes shape, that is, deformation of the object takes place. Usually the change in shape is so small that it cannot be detected by just watching the object. However, when very sensitive measuring instruments are used, very small changes in dimensions can be detected.

A force of attraction exists between all objects. The factors governing the size of this force F are the masses of the objects and the distances between their centres. Thus, if a person is taken as one object and the earth as a second object, a force of attraction exists between the person and the earth. This force is called the *gravitational force* and is the force that gives a person a certain weight when standing on the earth's surface. It is also this force that gives freely falling objects a constant acceleration in the absence of other forces.

Newton's laws of motion

To make a stationary object move or to change the direction in which the object is moving requires a force to be applied externally to the object. This concept is known as *Newton's first law of motion* and may be stated as:

An object remains in a state of rest, or continues in a state of uniform motion in a straight line, unless it is acted on by an externally applied force

Since a force is necessary to produce a change of motion, an object must have some resistance to a change in its motion. The force necessary to give a stationary pram a given acceleration is far less than the force necessary to give a stationary car the same acceleration. The resistance to a change in motion is called the *inertia* of an object and the amount of inertia depends on the mass of the object. Since a car has a much larger mass than a bicycle, the inertia of a car is much larger than that of a bicycle.

Newton's second law of motion may be stated as:

The acceleration of an object acted upon by an external force is proportional to the force and is in the same direction as the force

Thus, force α acceleration, or force = a constant x acceleration, this constant of proportionality being the mass of the object, i.e.

force = mass × acceleration

The unit of force is the newton (N) and is defined in terms of mass and acceleration. One newton is the force required to give a mass of 1 kilogram an acceleration of 1 metre per second squared. Thus

$F = ma$

where F is the force in newtons (N), m is the mass in kilograms (kg) and a is the acceleration in metres per second squared (ms^2), i.e.

$1\ N = 1\ kg\ m/s^2$

It follows that $1\ m/s^2 = 1\ N/kg$. Hence a gravitational acceleration of 9.81 m/s^2 is the same as a gravitational field of 9.81 N/kg.

Newton's third law of motion may be stated as:

For every force, there is an equal and opposite reacting force

Thus, an object on, say, a table, exerts a downward force on the table and the table exerts an equal upward force on the object, known as a *reaction force* or just a *reaction*.

Example 4.53

Calculate the force needed to accelerate a boat of mass 20 tonne uniformly from rest to a speed of 21.6 km/h in 10 minutes.

The mass of the boat, m, is 20 t, that is 20000 kg. The law of motion, $v = u + at$ can be used to determine the acceleration a. The initial velocity, u, is zero. The final velocity, v is 21.6 km/h, that is, 21.6/3.6 or 6 m/s. The time, t, is 10 min, that is, 600 s. Thus

$$6 = 0 + (a \times 600)\ \text{ or }\ a = \frac{6}{600} = \textbf{0.01 m/s}^2$$

From Newton's second law, $F = ma$, i.e.

Force = 20000 x 0.01 N = **200 N**

Test your knowledge 4.21

1 A lorry of mass 1350 kg accelerates uniformly from 9 km/h to reach a velocity of 45 km/h in 18 s. Determine (a) the acceleration of the lorry, (b) the uniform force needed to accelerate the lorry.

2 The tension in a rope lifting a crate vertically upwards is 2.8 kN. Determine its acceleration if the mass of the crate is 270 kg.

Example 4.54

The moving head of a machine tool requires a force of 1.2 N to bring it to rest in 0.8 s from a cutting speed of 30 m/min. Find the mass of the moving head.

From Newton's second law, $F = ma$, thus $m = F/a$, where force is given as 1.2 N. The law of motion $v = u + at$ can be used to find acceleration a, where $v = 0$, $u = 30$ m/min, that is 30/60 or 0.5 m/s, and $t = 0.8$ s. Thus,

$0 = 0.5 + (a \times 0.8)$

i.e. $\quad a = -\dfrac{0.5}{0.8} = -0.625$ m/s^2 (a *retardation* of 0.625 m/s^2)

Thus the mass, $m = 1.2/0.625 =$ **1.92 kg**

Activity 4.6

Investigate the battery starting system used in a conventional motor car. Identify the electrical and mechanical components used in the system, including the battery, solenoid, starter motor, alternator, rectifier, and regulator. Explain how each component works and how they work together. Estimate the energy required to start the car and how much energy is required to charge the battery. Comment on the efficiency of the system and present your findings in the form of a brief class presentation using appropriate diagrams and visual aids.

Activity 4.7

An aircraft manufacturing company has asked you to advise on a system that will be used to operate the flaps on a new light aircraft. The flaps are to be adjusted over a range extending from 0° to 40° when the pilot operates a control lever in the cockpit.

The company has considered the use of a mechanical system but is also keen to investigate the use of a motorized actuator. The company needs to know what factors should be taken into account when determining the power rating for the electric motors that will be used to drive the flaps.

Prepare a brief presentation to the design team explaining how the motorized flap system will work and how the power rating for the motors can be determined (include relevant formulae). You should also suggest advantages and disadvantages of the motorized system when compared with using a purely mechanical system.

Centripetal acceleration

When an object moves in a circular path at constant speed, its direction of motion is continually changing and hence its velocity (which depends on both magnitude *and direction*) is also continually changing. Since acceleration is the (change in velocity)/(time taken) the object has an acceleration.

If the tangential velocity is v and r is the radius of the circular path then the acceleration a is v^2/r and is towards the centre of the circle of motion. It is called the *centripetal acceleration*. If the mass of the rotating object is m, then by Newton's second law, the *centripetal force* is mv^2/r, and its direction is towards the centre of the circle of motion.

Example 4.55

An aircraft is turning at constant altitude, the turn following the arc of a circle of radius 1.5 km. If the maximum allowable acceleration of the aircraft is 2.5 g, determine the maximum speed of the turn in km/h. Take g as 9.8 m/s².

The acceleration of an object turning in a circle is v^2/r. Thus, to determine the maximum speed of turn, $v^2/r = 2.5$ g.

$$v = \sqrt{2.5gr} = \sqrt{2.5 \times 9.81 \times 1500} = \sqrt{36750}$$

$$= 191.7 \text{ m/s}$$

Now 1 m/s = 3.6 km/h, thus

191.7 m/s = 191.7 x 3.6 km/h = **690 km/h**

Angular motion

Linear velocity is defined as the rate of change of linear displacement s with respect to time t. Thus, for motion in a straight line (i.e. linear motion):

$$\text{linear velocity} = \frac{\text{change in distance}}{\text{change in time}}$$

i.e.

$$v = \frac{s}{t} \tag{4.1}$$

The unit of linear velocity is metres per second (m/s).

Angular velocity is defined as the rate of change of angular displacement θ, with respect to time t, and for an object rotating about a fixed axis at a constant speed:

$$\text{angular velocity} = \frac{\text{change in angle}}{\text{change in time}}$$

i.e.

$$\omega = \frac{\theta}{t} \tag{4.2}$$

The unit of angular velocity is radians per second (rad/s). Note that the speed of revolution of a wheel or a shaft is often specified in terms of revolutions per minute or revolutions per second but these units do not form part of a coherent system of units. The basis used in SI units is the angle turned through in one second. An object rotating at a constant speed of n revolutions per second subtends an angle of $2\pi n$ radians in one second, that is, its angular velocity:

$$\omega = 2\pi n \text{ rad/s} \tag{4.3}$$

$s = r\theta$ and from equation (4.2), $\theta = t$, hence $s = r \omega t$, or $s/t = r$.

However, from equation (4.1), $v = s/t$, hence

$$v = \omega r \qquad (4.4)$$

Equation (4.4) gives the relationship between linear velocity, v, and angular velocity, ω.

Example 4.56

A wheel of diameter 540 mm is rotating at $(1500/\pi)$ rev/min. Calculate the angular velocity of the wheel and the linear velocity of a point on the rim of the wheel.

From equation (4.3), angular velocity $\omega = 2\pi n$, where n is the speed of revolution in revolutions per second, i.e.

$n = 1500/60\pi$ revolutions per second. Thus

$$\omega = 2\pi\; \frac{1500}{60\pi} = \textbf{50 rad/s}$$

The linear velocity of a point on the rim, $v = r$, where r is the radius of the wheel, i.e. 0.54/2 or 0.27 m. Thus

$$v = 50 \times 0.27 = \textbf{13.5 m/s}$$

Test your knowledge 4.23

A pulley driving a belt has a diameter of 360 mm and is turning at 2700/π revolutions per minute. Find the angular velocity of the pulley and the linear velocity of the belt assuming that no slip occurs.

Linear acceleration, a, is defined as the rate of change of linear velocity with respect to time (as introduced earlier). For an object whose linear velocity is increasing uniformly:

$$\text{linear acceleration} = \frac{\text{change in linear velocity}}{\text{change in time}}$$

i.e.

$$\text{average acceleration, } a = \frac{v_2 - v_1}{t}\;\; \text{m/s}^2 \qquad (4.5)$$

The unit of linear acceleration is metres per second squared (m/s^2). Rewriting equation (4.5) with v_2 as the subject of the formula gives:

$$v_2 = v_1 + at \qquad (4.6)$$

Angular acceleration, α, is defined as the rate of change of angular velocity with respect to time. For an object whose angular velocity is increasing uniformly:

$$\text{angular acceleration} = \frac{\text{change in angular velocity}}{\text{change in time}}$$

that is,

$$\alpha = \frac{\omega_2 - \omega_1}{t} \qquad (4.7)$$

The unit of angular acceleration is radians per second squared (rad/s^2). Rewriting equation (4.7) with ω_2 as the subject of the formula gives:

$$\omega_2 = \omega_1 + \alpha t \tag{4.8}$$

From equation (4.8), $v = \omega\, r$. For motion in a circle having a constant radius r, $v_2 = \omega_2 r$ and $v_1 = \omega_1 r$, hence equation (4.5) can be rewritten as

$$a = \frac{\omega_2 r - \omega_1 r}{t} = \frac{r(\omega_2 - \omega_1)}{t}$$

But from equation (4.7),

$$\frac{\omega_2 - \omega_1}{t} = \alpha$$

Hence

$$a = r\alpha \tag{4.9}$$

Example 4.57

The speed of a shaft increases uniformly from 300 revolutions per minute to 800 revolutions per minute in 10 s. Find the angular acceleration, correct to four significant figures.

From equation 4.8, $\omega_2 = \omega_1 + \alpha t$, hence $\alpha = \dfrac{\omega_2 - \omega_1}{t}$

Initial angular velocity,

ω_1 = 300 revolutions per minute

$$= \frac{300}{60} \text{ revolutions per second} = \frac{300 \times 2\pi}{60} \text{ rad/s}$$

Final angular velocity,

$$\omega_2 = \frac{800 \times 2\pi}{60} \text{ rad/s}$$

Hence,

angular acceleration $\alpha = \dfrac{\left(\dfrac{800 \times 2\pi}{60} - \dfrac{300 \times 2\pi}{60}\right)}{10}$ rad/s^2

$$= \frac{500 \times 2\pi}{60 \times 10} = \textbf{5.236 rad/s}^2$$

Example 4.58

If the diameter of the shaft in Example 4.56 is 50 mm, determine the linear acceleration of the shaft, correct to three significant figures.

From equation 4.9, $a = r\alpha$

The shaft radius is 0.05/2 m and the angular acceleration α is

5.236 rad/s², thus the linear acceleration, $a = 0.05/2 \times 5.236 = \mathbf{0.131\ m/s^2}$

Further equations of motion

From equation (4.1), $s = vt$, and if the linear velocity is changing uniformly from v_1 to v_2, then s = mean linear velocity × time, i.e.

$$s = \frac{v_1 + v_2}{2}\ t \tag{4.10}$$

From equation (4.6), $\theta = \omega t$, and if the angular velocity is changing uniformly from ω_1 to ω_2, then θ = mean angular velocity × time, i.e.

$$\theta = \frac{\omega_2 + \omega_1}{2}\ t \tag{4.11}$$

Two further equations of linear motion may be derived from equations (4.6) and (4.9):

$$s = v_1 t + \tfrac{1}{2}\,at^2 \tag{4.12}$$

and

$$v_2^2 = v_1^2 + 2as \tag{4.13}$$

Two further equations of angular motion may be derived from equations (4.8) and (4.10):

$$\theta = \omega_1 t + \tfrac{1}{2}\,\alpha t^2 \tag{4.14}$$

and

$$\omega_2^2 = \omega_1^2 + 2\alpha\theta \tag{4.15}$$

Example 4.59

The shaft of an electric motor, initially at rest, accelerates uniformly for 0.4 s at 15 rad/s². Determine the angle turned through by the shaft, in radians, in this time.

From equation (4.14), $\theta = \omega_1 t + \tfrac{1}{2}\alpha t^2$. Since the shaft is initially at rest, $\omega_1 = 0$ and $\theta = \tfrac{1}{2}\alpha t^2$. The angular acceleration, α, is 15 rad/s² and time t is 0.4 s, hence angle turned through, $\theta = \tfrac{1}{2} \times 15 \times 0.4^2 = \mathbf{1.2\ rad}$

Activity 4.8

Earlier, we mentioned that if a body moves as a result of a force being applied to it, the force is said to do work on the body. In a rotary system the amount of work done is the product of the applied torque (T) and the angle turned (θ). Show that the rate of doing work in a rotary system is given by $T\omega$ where T is the applied torque and ω is the angular velocity. Present your answer in the form of a class handout.

Test your knowledge 4.24

A disk accelerates uniformly from 300 revolutions per minute to 600 revolutions per minute in 25 s. Determine (a) its angular acceleration and (b) the linear acceleration of a point on the rim of the disk, if the radius of the disk is 250 mm. (c) Calculate the number of revolutions the disk makes during the 25 s acceleration period.

Activity 4.9

Investigate the operation of a typical rechargeable power drill suitable for the home DIY enthusiast. Produce a full specification for the product and explain the meaning of each rating. Explain the relationships that exists between (a) torque and speed and (b) supply current and power. Present your findings in the form of a word processed fact sheet.

Review questions

1 Determine the density of an alloy material if a 10 cm^3 sample of the material has a mass of 45 g.

2 A tank contains 50 litres of petrol having a density of 700 gm/m^2. What will be the mass of petrol stored in the tank?

3 The jib of an overhead crane moves from A to B through a distance of 10 m when a force of 5 kN is applied to it. Calculate the work done moving the jib.

4 If it takes 25 s to move the crane jib in question 3, determine the mechanical power required.

5 A lift exerts a force of 900 N when raising a load through a height of 10 m. If the mechanical efficiency of the hoist is 60% and the efficiency of the motor drive is 50% determine the electrical energy required to operate the lift.

6 Define the terms (a) potential energy and (b) kinetic energy.

7 Briefly explain the principle of conservation of energy.

8 A missile having a mass of 15 kg is travelling at a constant speed of 50 m/s. Determine the kinetic energy stored in the missile.

9 A drop hammer has a mass of 500 kg and it operates from a height of 4 m. Determine the kinetic energy stored in the drop hammer.

10   A power supply is rated at 24 V, 400 mA. What value of load resistor is required to test the power supply at full load and what is the minimum power rating for the resistor used?

11   A heater operates from a 12 V supply and consumes 4 A. If the heater operates for a continuous period of 10 hours how much energy is supplied?

12   Define the terms:
(a) specific heat capacity
(b) latent heat of fusion
(c) latent heat of vaporization.

13   Determine the heat energy required to convert 1.5 kg of ice at 0°C to water at 0°C.

14   Explain, with reference to the atomic structure of a material, why some materials are conductors and some are insulators.

15   An overhead power cable has a length of 2.5 km and a cross sectional area of 50 mm^2. Determine the voltage dropped along the length of the cable if a current of 10 A flows in it and the material has a resistivity of 0.025×10^{-6} Ωm.

16   Determine the power loss in question 15.

17   Four 10 Ω resistors are available. Sketch circuit diagrams showing how these resistors can be connected in order to produce resistances of: (a) 40 Ω, (b) 2.5 Ω, (c) 13.33 Ω, and (d) 25 Ω.

18   State (a) Kirchhoff's current law and (b) Kirchhoff's voltage law.

19   A sinusoidal voltage has an r.m.s. value of 110 V and a frequency of 60 Hz. Write down an expression for the instantaneous voltage.

20   In question 19, determine the voltage at (a) $t = 3.5$ ms and (b) $t = 15$ ms.

21   An aircraft is travelling at 250 km/h along a straight path. After some time, additional engine thrust is applied such that the aircraft accelerates at a constant rate of 1.2 m/s^2 for a period of 2 minutes. Determine the final velocity reached.

22   An engine drive shaft rotates at a constant speed of 3600 revolutions per minute. If the shaft speed increases to 4800 revolutions per minute in a time of 30 s, determine the angular acceleration.

Unit 5 | Applied mathematics in engineering

Summary

This section aims to develop in the reader an understanding of the mathematical skills essential to engineering. The material is divided into four main sections; algebra, equations, trigonometry and graphs. To help students develop skills in mathematics a number of typical engineering problem solving activities have also been included.

Notation

Indices

The lowest factors of 2000 are $2 \times 2 \times 2 \times 2 \times 5 \times 5 \times 5$. These factors are written as $2^4 \times 5^3$, where 2 and 5 are called *bases* and the numbers 4 and 3 are called *indices*.

When an index is an integer it is called a *power*. Thus, 2^4 is called 'two to the power of four', and has a base of 2 and an index of 4. Similarly, 5^3 is called 'five to the power of 3' and has a base of 5 and an index of 3.

Special names may be used when the indices are 2 and 3, these being called 'squared' and 'cubed', respectively. Thus 7^2 is called 'seven squared' and 9^3 is called 'nine cubed'. When no index is shown, the power is 1, i.e. 2^1 means 2.

Reciprocal

The *reciprocal* of a number is when the index is -1 and its value is given by 1 divided by the base. Thus the reciprocal of 2 is 2^{-1} and its value is ½ or 0.5. Similarly, the reciprocal of 4 is 4^{-1} which means ¼ or 0.25.

Square root

The *square root* of a number is when the index is ½, and the square root of 2 is written as $2^{½}$ or $\sqrt{2}$. The value of a square root is the value of the base which when multiplied by itself gives the number. Since $3 \times 3 = 9$, then $\sqrt{9} = 3$. However, $(-3) \times (-3) = 9$, so $\sqrt{9} = -3$. There are always two answers when finding the square root of a number and this is shown by putting both a + and a − sign in front of the answer to a square root problem. Thus $\sqrt{9} = \pm 3$ and $4^{\frac{1}{2}} = \sqrt{4} = \pm 2$, and so on.

Laws of indices

When simplifying calculations involving indices, certain basic rules or laws can be applied, called the *laws of indices*. These are given below:

(i)   When multiplying two or more numbers having the same base, the indices are added. Thus $3^2 \times 3^4 = 3^{2+4} = 3^6$.

(ii) When a number is divided by a number having the same base, the indices are subtracted. Thus $3^5/3^2 = 3^{5-2} = 3^3$.

(iii) When a number which is raised to a power is raised to a further power, the indices are multiplied. Thus $(3^5)^2 = 3^{5 \times 2} = 3^{10}$.

(iv) When a number has an index of 0, its value is 1. Thus $3^0 = 1$.

(v)   A number raised to a negative power is the reciprocal of that number raised to a positive power. Thus $3^{-4} = 1/3^4$. Similarly, $1/2^{-3} = 2^3$.

(vi) When a number is raised to a fractional power the denominator of the fraction is the root of the number and the numerator is the power. Thus

$$8^{\frac{2}{3}} = \sqrt[3]{8^2} = (2)^2 = 4$$

and

$$25^{\frac{1}{2}} = \sqrt{25^1} = \pm 5$$

Example 5.1

Evaluate: (a) $5^2 \times 5^3$, (b) $3^2 \times 3^4 \times 3$ and (c) $2 \times 2^2 \times 2^5$

From law (i):

(a) $5^2 \times 5^3 = 5^{(2+3)} = 5^5 = 5 \times 5 \times 5 \times 5 \times 5 = \mathbf{3125}$

(b) $3^2 \times 3^4 \times 3 = 3^{(2+4+1)} = 3^7 = 3 \times 3 \times \dots$ to 7 terms $= \mathbf{2187}$

(c) $2 \times 2^2 \times 2^5 = 2^{(1+2+5)} = 2^8 = \mathbf{256}$

Example 5.2

Find the value of: (a) $7^5/7^3$ (b) $5^7/5^4$

From law (ii):

(a) $\dfrac{7^5}{7^3} = 7^{(5-3)} = 7^2 = \mathbf{49}$

(b) $\dfrac{5^7}{5^4} = 5^{(7-4)} = 5^3 = \mathbf{125}$

Example 5.3

Evaluate: (a) $5^2 \times 5^3 \div 5^4$ and (b) $(3 \times 3^5) \div (3^2 \times 3^3)$

From laws (i) and (ii):

(a) $5^2 \times 5^3 \div 5^4 = \dfrac{5^2 \times 5^3}{5^4} = \dfrac{5^{(2+3)}}{5^4}$

$\qquad\qquad\qquad = \dfrac{5^5}{5^3} = 5^{(5-4)} = 5^1 = \mathbf{5}$

(b) $(3 \times 3^5) \div (3^2 \times 3^3) = \dfrac{3 \times 3^5}{3^2 \times 3^3} = \dfrac{3^{(1+5)}}{3^{(2+3)}}$

$\qquad\qquad\qquad\qquad = \dfrac{3^6}{3^5} = 3^{(6-5)} = 3^1 = \mathbf{3}$

Example 5.4

Simplify: (a) $(2^3)^4$ and (b) $(3^2)^5$, expressing the answers in index form.

From law (iii):

(a) $(2^3)^4 = 2^{3 \times 4} = \mathbf{2^{12}}$

(b) $(3^2)^5 = 3^{2 \times 5} = \mathbf{3^{10}}$

Example 5.5

Evaluate: $\dfrac{(10^2)^3}{10^4 \times 10^2}$.

From the laws of indices:

$\dfrac{(10^2)^3}{10^4 \times 10^2} = \dfrac{10^{(2 \times 3)}}{10^{(4+2)}} = \dfrac{10^6}{10^6} = 10^{6-6} = 10^0 = \mathbf{1}$

Example 5.6

Evaluate (a) $4^{1/2}$ (b) $16^{3/4}$ (c) $27^{2/3}$ (d) $9^{-1/2}$

(a) $4^{1/2} = \sqrt{4} = \mathbf{\pm 2}$

(b) $16^{3/4} = \sqrt[4]{16^3} = (2)^3 = \mathbf{8}$

(Note that it does not matter whether the 4th root of 16 is found first or whether 16 cubed is found first – the same answer will result.)

(c) $27^{2/3} = \sqrt[3]{27^2} = (3)^2 = \mathbf{9}$

(d) $9^{-1/2} = \dfrac{1}{9^{\frac{1}{2}}} = \dfrac{1}{\sqrt{9}} = \dfrac{1}{\pm 3} = \mathbf{\pm 1/3}$

Test your knowledge 5.1

In questions 1 to 3, simplify the expressions given, expressing the answers in index form.

1 (a) $3^3 \times 3^4$ (b) $4^2 \times 4^3 \times 4^4$

2 (a) $\dfrac{2^4}{2^3}$ (b) $\dfrac{3^7}{3^2}$

3 (a) $(7^2)^3$ (b) $(3^3)^2$

4 Evaluate:

(a) $\dfrac{1^{-1}}{3^2}$ (b) $81^{0.25}$

(c) $16^{-1/4}$ (d) $\dfrac{4^{1/2}}{9}$

Standard form

A number written with one digit to the left of the decimal point and multiplied by 10 raised to some power is said to be written in *standard form*. Thus: 5837 is written as 5.837×10^3 in standard form, and 0.0415 is written as 4.15×10^{-2} in standard form.

When a number is written in standard form, the first factor is called the *mantissa* and the second factor is called the *exponent*. Thus the number 5.8×10^3 has a mantissa of 5.8 and an exponent of 10^3.

(i) Numbers having the same exponent can be added or subtracted in standard form by adding or subtracting the mantissae and keeping the exponent the same. Thus:

$$2.3 \times 10^4 + 3.7 \times 10^4 = (2.3 + 3.7) \times 10^4 = 6.0 \times 10^4,$$

and

$$5.9 \times 10^{-2} - 4.6 \times 10^{-2} = (5.9 - 4.6) \times 10^{-2}$$
$$= 1.3 \times 10^{-2}$$

When numbers have different exponents, one way of adding or subtracting the numbers is to express one of the numbers in non-standard form, so that both numbers have the same exponent. Thus:

$$2.3 \times 10^4 + 3.7 \times 10^3 = 2.3 \times 10^4 + 0.37 \times 10^4$$
$$= (2.3 + 0.37) \times 10^4$$
$$= 2.67 \times 10^4$$

Alternatively,

$$2.3 \times 10^4 + 3.7 \times 10^3 = 23000 + 3700 = 26700 = 2.67 \times 10^4$$

(ii) The laws of indices are used when multiplying or dividing numbers given in standard form. For example,

$$(22.5 \times 10^3) \times (5 \times 10^2) = (2.5 \times 5) \times (10^{3+2})$$
$$= 12.5 \times 10^5 \text{ or } 1.25 \times 10^6$$

Similarly,

$$\frac{6 \times 10^6}{1.5 \times 10^2} = \frac{6}{1.5} \times 10^{4-2} = 4 \times 10^2$$

$$\frac{150 \times 10^{-6}}{30 \times 10^3} = \frac{5}{1} \times 10^{-6-3} = 5 \times 10^{-9}$$

and,

$$\frac{9 \times 10^3}{54 \times 10^9} = \frac{1}{6} \times 10^{3-9} = 0.1667 \times 10^{-6} = 1.667 \times 10^{-7}$$

Note that in the last example we have multiplied the number by 10 and reduced the index by 1 (i.e. from −6 to −7).

Test your knowledge 5.2

In questions 1 and 2, express in standard form

1 (a) 73.9 (b) 1128.4 (c) 197.62

2 (a) 0.2401   (b) 0.0174

   (c) 0.00923

In questions 3 and 4, evaluate in standard form

3 (a) $4.831 \times 10^2 + 1.24 \times 10^3$

   (b) $3.24 \times 10^{-3} - 1.11 \times 10^{-4}$

4 (a) $\dfrac{6 \times 10^{-3}}{3 \times 10^{-5}}$

   (b) $\dfrac{(2.4 \times 10^3) \times (3 \times 10^{-2})}{(4.8 \times 10^4)}$

   (c) $\dfrac{(2.4 \times 10^3) \times (3 \times 10^{-2})}{(4.8 \times 10^4)}$

Example 5.7

Express in standard form: (a) 38.71 (b) 3746 (c) 0.0124

For a number to be in standard form, it is expressed with only one digit to the left of the decimal point. Thus:

(a) 38.71 must be divided by 10 to achieve one digit to the left of the decimal point and it must also be multiplied by 10 to maintain the equality, i.e.

$$38.71 = \frac{38.71}{10} \times 10 = \mathbf{3.871 \times 10^1} \text{ in standard form}$$

(b) $$3746 = \frac{3746}{1000} \times 1000 = \mathbf{3.746 \times 10^3} \text{ in standard form}$$

(c) $$0.0124 = 0.0124 \times \frac{100}{100} = \frac{1.24}{100} = \mathbf{1.24 \times 10^{-2}} \text{ in standard form}$$

Algebra

Algebra is that part of mathematics in which the relations and properties of numbers are investigated by means of general symbols. For example, the area of a rectangle is found by multiplying the length by the breadth; this is expressed algebraically as $A = l \times b$, where A represents the area, l the length and b the breadth.

The basic laws introduced in arithmetic are generalized in algebra. Let a, b, c and d represent any four numbers. Then:

(i) $a + (b + c) = (a + b) + c$

(ii) $a(bc) = (ab)c$

(iii) $a + b = b + a$

(iv) $ab = ba$

(v) $a(b + c) = ab + ac$

(vi) $\dfrac{a+b}{c} = \dfrac{a}{c} + \dfrac{b}{c}$

(vii) $(a + b)(c + d) = ac + ad + bc + bd$

Example 5.8

Evaluate $3ab - 2bc + abc$ when $a = 1$, $b = 3$ and $c = 5$

Replacing a, b and c with their numerical values gives:

$3ab - 2bc + abc = (3 \times 1 \times 3) - (2 \times 3 \times 5) + (1 \times 3 \times 5)$

$$= 9 - 30 + 15 = \mathbf{-6}$$

Example 5.9

Find the sum of $3x$, $2x$, $-x$ and $-7x$

The sum of the positive terms is $3x + 2x = 5x$

The sum of the negative terms is $x + 7x = 8x$

Taking the sum of the negative terms from the sum of the positive terms gives:

$5x - 8x = -3x$

Alternatively

$3x + 3x + (-x) + (-7x) = 3x + 2x - x - 7x = -3x$

Example 5.10

Find the sum of $4a$, $3b$, c, $-2a$, $-5b$ and $6c$

Each symbol must be dealt with individually.

for the 'a' terms: $+4a - 2a = 2a$

for the 'b' terms: $+3b - 5b = -2b$

for the 'c' terms: $+c + 6c = 7c$

Thus $4a + 3b + c + (-2a) + (-5b) + 6c = 4a + 3b + c - 2a - 5b + 6c$

$= 2a - 2b + 7c$

Laws of indices

Expressed algebraically, the laws of indices are:

(i) $a^m \times a^n = a^{m+n}$

(ii) $a^m \div a^n = a^{m-n}$

(iii) $(a^m)^n = a^{mn}$

(iv) $a^{m/n} = \sqrt[n]{a^m}$

(v) $a^{-n} = 1/a^n$

(vi) $a^0 = 1$

(vii) $a^1 = a$

Example 5.11

Simplify $a^3 b^2 c \times ab^3 c^5$

Grouping like terms gives: $a^3 \times a \times b^2 \times b^3 \times c \times c^5$

Using the first law of indices gives: $a^{3+1} \times b^{2+3} \times c^{1+5}$

i.e. $a^4 \times b^5 \times c^6 = a^4 b^5 c^6$

Example 5.12

Simplify $a^3b^2c^4 \div abc^{-2}$ and evaluate when $a = 3$, $b = \frac{1}{8}$ and $c = 2$

Using the second law of indices,

$$\frac{a^3}{a} = a^{3-1} = a^2, \frac{b^2}{b} = b^{2-1} = b \text{ and } \frac{c^4}{c^{-2}} = c^{4--2} = c^{4+2} = c^6$$

Thus $\dfrac{a^3b^2c^4}{abc^{-2}} = \boldsymbol{a^2bc^6}$

When $a = 3$, $b = \frac{1}{8}$, and $c = 2$,

$$a^2bc^6 = (3)^2 \times \tfrac{1}{8} \times (2)^6 = \frac{9 \times 64}{8} = \boldsymbol{72}$$

Example 5.13

Simplify $\dfrac{x^2y^3 + xy^2}{xy}$

Algebraic expressions of the form $\dfrac{a+b}{c}$ can be split into $\dfrac{a}{c} + \dfrac{b}{c}$

Thus

$$\frac{x^2y^3 + xy^2}{xy} = \frac{x^2y^3}{xy} + \frac{xy^2}{xy} = x^{2-1}y^{3-1} + x^{1-1}y^{2-1} = \boldsymbol{xy^2 + y}$$

(since $x^0 = 1$, from the sixth law of indices).

Example 5.14

Simplify $(p^3)^{1/2}(q^2)^4$

Using the third law of indices gives:

$$p^{3 \times (1/2)}q^{2 \times 4} = \boldsymbol{p^{(3/2)}q^8}$$

Test your knowledge 5.3

1 Simplify $\dfrac{(x^2y^3z)\,(x^3y^4z^2)}{(xyz)}$

and evaluate when $x = 2$, $y = -\frac{1}{2}$ and $x = 3$

2 Simplify $\dfrac{a^3b + a^2b^3}{a^2b^2}$

3 Simplify $\dfrac{(a^3\sqrt{b}\ c^{-2})\,(ab)^{-\frac{1}{2}}}{(\sqrt{a})\,(c^{-3})}$

Brackets and factorization

When two or more terms in an algebraic expression contain a common factor, then this factor can be shown outside of a bracket. For example

$$ab + ac = a(b + c)$$

which is simply the reverse of law (v) of indices, and

$$6px + 2py - 4pz = 2p(3x + y - 2z)$$

This process is called *factorization*.

Example 5.15

Remove the brackets and simplify the expression

$(3a + b) + 2(b + c) - 4(c + d)$

Both b and c in the second bracket have to be multiplied by 2, and c and d in the third bracket by -4 when the brackets are removed. Thus:

$(3a + b) + 2(b + c) - 4(c + d) = 3a + b + 2b + 2c - 4c - 4d$

Collecting similar terms together gives: **$3a + 3b - 2c - 4d$**

Example 5.16

Simplify $a^2 - (2a - ab) - a(3b + a)$

When the brackets are removed, both $2a$ and $-ab$ in the first bracket must be multiplied by -1 and both $3b$ and a in the second bracket by $-a$. Thus

$a^2 - (2a - ab) - a(3b + a) = a^2 - 2a + ab - 3ab - a^2$

Collecting similar terms together gives: $-2a - 2ab$

Since $-2a$ is a common factor the answer can be expressed as **$-2a(1 + b)$**

Example 5.17

Simplify $(a + b)(a - b)$

Each term in the second bracket has to be multiplied by each term in the first bracket. Thus:

$(a + b)(a - b) = a(a - b) + b(a - b) = a^2 - ab + ab - b^2 = \mathbf{a^2 - b^2}$

Example 5.18

Remove the brackets from the expression $(x - 2y)(3x + y^2)$

$(x - 2y)(3x + y^2) = x(3x + y^2) - 2y(3x + y^2)$

$= \mathbf{3x^2 + xy^2 - 6xy - 2y^3}$

Example 5.19

Simplify $(3x - 3y)^2$

$(2x - 3y)^2 = (2x - 3y)(2x - 3y)$

$= 2x(2x - 3y) - 3y(2x - 3y)$

$= 4x^2 - 6xy - 6xy + 9y^2$

$= \mathbf{4x^2 - 12xy + 9y^2}$

Example 5.20

Remove the brackets from the expression $2[p^2 - 3(q + r) + q^2]$

In this example there are two brackets and the 'inner' one is removed first.

Hence $2[p^2 - 3(q + r) + q^2] = 2[p^2 - 3q - 3r + q^2]$

$$= 2p^2 - 6q - 6r + 2q^2$$

Example 5.21

Remove the brackets and simplify the expression

$2a - [3\{2(4a - b) - 5(a + 2b)\} + 4a]$

Removing the innermost brackets gives:

$$2a - [3\{8a - 2b - 5a - 10b\} + 4a]$$

Collecting together similar terms gives:

$$2a - 3[\{3a - 12b\} + 4a]$$

Removing the 'curly' brackets gives:

$$2a - [9a - 36b + 4a]$$

Collecting together similar terms gives:

$$2a - [13a - 36b]$$

Removing the outer brackets gives:

$$2a - 13a + 36b$$

i.e.

$$-11a + 36b = 36b - 11a$$

Example 5.22

Factorize (a) xy - 3xz, (b) $4a^2 + 16ab^3$, (c) $3a^2b - 6ab^2 + 15ab$

For each part of this problem, the *highest common factor* (HCF) of the terms will become one of the factors. Thus:

(a) $xy - 3xz = x(y - 3z)$

(b) $4a^2 + 16ab^3 = 4a(a + 4b^3)$

(c) $3a^2b - 6ab^2 + 15ab = 3ab(a - 2b + 5)$

Test your knowledge 5.4

1 Remove the brackets and simplify:

(a) $3(x - y) - 2(2y - x)$

(b) $(2a - b)(a + b)$

(c) $3 - 4\{x(x - 2y) - (x - y)^2\}$

2 Factorize:

(a) $14ab^2 - 35ab$

(b) $ax - ay + bx - by$

Fundamental laws and precedence

The *laws of precedence* that apply to arithmetic also apply to algebraic expressions. The order is Brackets, Of, Division, Multiplication, Addition and Subtraction (i.e. BODMAS).

Example 5.23

Simplify $2a + 5a \times 3a - a$

Multiplication is performed before addition and subtraction thus:

$$2a + 5a \times 3a - a = 2a + 15a^2 - a$$
$$= a + 15a^2 = \mathbf{a(1 + 15a)}$$

Example 5.24

Simplify $(a + 5a) \times 2a - 3a$

The order of precedence is brackets, multiplication, then subtraction. Hence

$$(a + 5a) \times 2a - 3a = 6a \times 2a - 3a = 12a^2 - 3a$$
$$= \mathbf{3a(4a - 1)}$$

Example 5.25

Simplify $a + 5a \times (2a - 3a)$

The order of precedence is brackets, multiplication, then subtraction. Hence

$$a + 5a \times (2a - 3a) = a + 5a \times -a = a + -5a^2$$
$$= a - 5a^2 = \mathbf{a(1 - 5a)}$$

Example 5.26

Simplify $a \div 5a + 2a - 3a$

The order of precedence is division, then addition and subtraction. Hence

$$a \div 5a + 2a - 3a = \frac{a}{5a} + 2a - 3a = \frac{1}{5} + 2a - 3a$$
$$= \mathbf{0.2 - a}$$

Test your knowledge 5.5

Simplify the following:

1 $(3a - a) \div 2a - a$

2 $3b + 2 \div 3b + 2 \times 4 - 5b$

3 $2t^2 - (4st)(2t) \div 4s + st$

4 $1/3$ of $3p + 4p \, (3p - p)$

Example 5.27

Simplify $(3c + 2c)(4c + c) \div (5c - 8c)$

The order of precedence is brackets, division and multiplication. Hence:

$$(3c + 2c)(4c + c) \div (5c - 8c) = 5c \times 5c \div -3c$$
$$= 5c \times \frac{5c}{-3c}$$
$$= 5c \times \frac{5c}{-3c} = 5c \times \frac{5}{-3} = \mathbf{-25/3 \, c}$$

Equations

Expressions, identities and equations

$(3x - 5)$ is an example of an *algebraic expression*, whereas $3x - 5 = 1$ is an example of an *equation* (i.e. it contains an 'equals' sign).

An equation is simply a statement that two quantities are equal. For example,

$$1 \text{ m} = 1000 \text{ mm} \quad \text{or} \quad y = mx + c.$$

An *identity* is a relationship which is true for all values of the unknown, whereas an equation is only true for particular values of the unknown. For example, $3x - 5 = 1$ is an equation, since it is only true when $x = 2$, whereas $3x \equiv 8x - 5x$ is an identity since it is true for all values of x. (Note that '$\equiv$' means 'is identical to'.)

Simple linear equations (or equations of the first degree) are those in which an unknown quantity is raised only to the power 1.

To '*solve an equation*' means 'to find the value of the unknown'. Any arithmetic operation may be applied to an equation *as long as the equality of the equation is maintained.*

Example 5.28

Solve the equation $4x = 20$

Dividing each side of the equation by 4 gives: $4x/4 = 20/4$

(Note that the same operation has been applied to both the left-hand side (LHS) and the right-hand side (RHS) of the equation.)

Cancelling gives $x = 5$, which is the solution to the equation.

Solutions to simple equations should always be checked and this is accomplished by substituting the solution into the original equation. In this case,

LHS = 4(5) = 20 = RHS.

Example 5.29

Solve $\dfrac{2x}{5} = 6$

The LHS is a fraction and this can be removed by multiplying both sides of the equation by 5. Hence

$$5 \times \frac{2x}{5} = 5 \times 6$$

Cancelling gives: $2x = 30$

Dividing both sides of the equation by 2 gives:

$$\frac{2x}{2} = \frac{30}{2} \quad \text{i.e. } x = 15$$

Example 5.30

Solve $a - 5 = 8$

Adding 5 to both sides of the equation gives:

$$a - 5 + 5 = 8 + 5$$

i.e. $a = 13$

The result of the above procedure is to move the '−5' from the LHS of the original equation, across the equals sign, to the RHS, but the sign is changed to +.

Example 5.31

Solve $6x + 1 = 2x + 9$

In such equations the terms containing x are grouped on one side of the equation and the remaining terms grouped on the other side of the equation.

As in Example 5.30, changing from one side of an equation to the other must be accompanied by a change of sign.

Thus since $6x + 1 = 2x + 9$

then $6x - 2x = 9 - 1$

$$4x = 8$$

$$4x/4 = 8/4$$

i.e. $x = 2$

Check:

LHS of original equation $= 6(2) + 1 = 13$

RHS of original equation $= 2(2) + 9 = 13$

Hence the solution $x = 2$ is correct.

Example 5.32

Solve $3(x - 2) = 9$

Removing the bracket gives: $3x - 6 = 9$

Rearranging gives: $3x = 9 + 6$

$$3x = 15$$

$$3x/3 = 15/3$$

i.e. $x = 5$

Check:

LHS $= 3(5 - 2) = 3(3) = 9 =$ RHS

Hence the solution $x = 5$ is correct.

Example 5.33

Solve $4(2r - 3) - 2(r - 4) = 3(r - 3) - 1$

Removing brackets gives: $8r - 12 - 2r + 8 = 3r - 9 - 1$

Rearranging gives: $\qquad 8r - 2r - 3r = -9 - 1 + 12 - 8$

i.e. $\qquad\qquad\qquad 3r = -6$

$$3r/3 = -6/3$$

$$\boldsymbol{r = -2}$$

Check: LHS $= 4(-4 - 3) - 2(-2 - 4) = -28 + 12 = -16$

$\qquad$ RHS $= 3(-2 - 3) - 1 = -15 - 1 = -16$

Hence the solution $r = -2$ is correct.

Example 5.34

Solve $3/x = 4/5$

The lowest common multiple (LCM) of the denominators, i.e. the lowest algebraic expression that both x and 5 will divide into, is $5x$. Multiplying both sides by $5x$ gives:

$$5x \times \frac{3}{x} = 5x \times \frac{4}{5}$$

Cancelling gives $\quad 15 = 4x \quad$ or $\quad 4x = 15$

Thus $\quad x = \dfrac{15}{4} \quad$ i.e. $\boldsymbol{x = 3\frac{3}{4}}$

(Note that when there is only one fraction on each side of an equation, 'cross-multiplication' can be applied. In this example, if

$$\frac{3}{x} = \frac{4}{5}$$

then $\quad 3 \times 5 = 4 \times x \quad$ which is a quicker way of arriving at the solution.)

Example 5.35

Solve $2\sqrt{d} = 8$

To avoid possible errors it is usually best to arrange the term containing the square root on its own. Thus

$$\frac{2\sqrt{d}}{2} = \frac{8}{2} \qquad \text{i.e. } \sqrt{d} = 4$$

Squaring both sides gives: $\boldsymbol{d = 16}$, which may be checked in the original equation.

Test your knowledge 5.6

Solve the following equations:

1   $x + 3 = 7$

2   $4 - 3p = 2p - 11$

3   $\dfrac{2y}{5} + \dfrac{3}{4} + 5 = \dfrac{1}{20} - \dfrac{3y}{2}$

4   $\dfrac{\sqrt{b} + 3}{\sqrt{b}} = 2$

5   $8 = \dfrac{y^2}{2}$

Example 5.36

Solve $x^2 = 25$

This example involves a square term and thus is not a simple equation (it is, in fact, a quadratic equation). However the solution of such an equation is often required and is therefore included for completeness.

Whenever a square of the unknown is involved, the square root of both sides of the equation is taken. Hence

$$\sqrt{x^2} = \sqrt{25}$$

i.e. $x = 5$

However, $x = -5$ is also a solution of the equation because

$$(-5) \times (-5) = +25$$

Therefore, whenever the square root of a number is required there are always two answers, one positive, the other negative.

The solution of $x^2 = 25$ is thus written as $x = \pm 5$

Problems involving simple equations

Formulae are often used to solve practical engineering problems. The examples that follow will show you how to determine values from simple equations:

Example 5.37

The temperature coefficient of resistance α may be calculated from the formula $R_t = R_0(1 + \alpha t)$. Find α given that $R_t = 0.928$, $R_0 = 0.8$ and $t = 40°C$.

With problems like this it's a good idea to start by listing what you know:

$R_t = 0.928\ \Omega$

$R_0 = 0.8\ \Omega$

$t = 40°C$

$\alpha = ?$

Since $R_t = R_0(1 + \alpha t)$ then $0.928 = 0.8[1 + \alpha(40)]$

$$0.928 = (0.8 \times 1) + (0.8 \times \alpha \times 40)$$
$$0.928 = 0.8 + 32\alpha$$
$$0.928 - 0.8 = 32\alpha$$
$$0.128 = 32\alpha$$

$$\alpha = \frac{0.128}{32} = 0.004$$

The units of α are $\Omega / \Omega /°C$ or just $/°C$

Thus $\alpha = \mathbf{0.004\ per\ °C}$

Example 5.38

A copper wire has a length l of 1.5 km, a resistance R of 5 Ω and a resistivity ρ of 17.2 x 10⁻⁶ Ω mm. Find the cross-sectional area, a, of the wire, given that $R = \rho\,l/a$.

Once again, it is worth getting into the habit of summarizing what you know from the question and what you need to find (don't forget to include the units):

$R = 5\,\Omega$

$\rho = 17.2 \times 10^{-6}\,\Omega$ mm

$l = 1500 \times 10^{3}$ mm

$a = ?$

Since $R = \rho\,l/a$ then

$$5 = \frac{(17.2\times10^{-6})\,(1500\times10^{3})}{a}$$

Cross multiplying (i.e. exchanging the '5' for the 'a') gives:

$$a = \frac{(17.2\times10^{-6})\,(1500\times10^{3})}{5}$$

Now group the numbers and the powers of 10 as shown below:

$$a = \frac{17.2\times1500\times10^{-6}\times10^{3}}{5}$$

Next simplify as far as possible:

$$a = \frac{17.2\times1500}{5}\times10^{-6+3}$$

Finally, evaluate the result using your calculator:

$$a = 5160 \times 10^{-3} = 5.16$$

Since we have been working in mm, the result, a, will be in mm². Hence

$a = 5.16$ mm²

It's worth noting from the previous example that we have used the laws of indices to simplify the powers of ten before attempting to use the calculator to determine the final result. The alternative to doing this is to make use of the exponent facility on your calculator. Whichever technique you use it's important to be confident that you are correctly using the exponent notation. It's not unknown for students to produce answers that are incorrect by a factor of 1000 or even 1000000 – in an engineering context an error of this magnitude could be totally disastrous!

Example 5.39

The distance s metres travelled in time t seconds is given by the formula $s = ut + \frac{1}{2}at^2$, where u is the initial velocity in m/s and a is the acceleration in m/s². Find the acceleration of the body if it travels 168 m in 6 s, with an initial velocity of 10 m/s.

Now:

$s = 168$ m

$t = 6$ s

$u = 10$ m/s

$a = ?$

From $\quad s = ut + \frac{1}{2}at^2$

$\qquad 168 = (10 \times 6) + \frac{1}{2}\, a\, (6)^2$

$\qquad 168 = 60 + (\frac{1}{2} \times a \times 36)$

$\qquad 168 = 60 + 18a$

$\quad 168 - 60 = 18a$

$\qquad 108 = 18a$

$\qquad a = \dfrac{108}{18}$

Since we are working in metres and seconds, the units of a will be m/s² thus

$\qquad$ **$a = 6$ m/s²**

Activity 5.1

A light aircraft manufacturer has asked you to advise on the design of a new type of hang-glider in which a 2.5 m solid aluminium tie bar is used to reinforce the wings of the aircraft. To reduce weight, the manufacturer wishes to keep the cross–sectional area of the tie bar as small as possible, consistent with keeping the extension of the tie bar to less than 10 mm.

The manufacturer has asked you to determine the minimum cross-sectional area (in mm²) that can be used for the tie bar if it has to support a load of 2400 N.

The *modulus of elasticity*, E, of the aluminium tie bar is related to its length l (m), extension x (m) and cross-sectional area A (m²) when carrying a load of F (N) is given by:

$$E = \frac{Fl}{Ax}$$

Present your result in the form of a fully worked solution.

Simultaneous equations

Only one equation is necessary when finding the value of a *single unknown quantity* (as with simple equations). When an equation contains *two unknown quantities* it has an infinite number of solutions. When two equations are available connecting the same two unknown values then a unique solution is possible. Similarly, for three unknown quantities it is necessary to have three equations in order to solve for a particular value of each of the unknown quantities, and so on. Equations that have to be solved together to find the unique values of the unknown quantities, which are true for each of the equations, are called *simultaneous equations*.

There are two methods of solving simultaneous equations analytically:

(a) by *substitution*, and (b) by *elimination*.

The first method involves making one of the two unknowns the subject of one of the two formulae and then using the RHS of the equation in the second equation. The second method involves adding or subtracting the two equations in order to remove one of the two unknowns. The following examples show how this is done:

Example 5.40

Solve the following equations for x and y by substitution.

$$x + 2y = -1 \qquad (1)$$

$$4x - 3y = 18 \qquad (2)$$

From equation (1): $x = -1 - 2y$

Substituting this expression for x into equation (2) gives:

$$4(-1 - 2y) - 3y = 18$$

This is not a simple equation in y.

Removing the bracket gives:

$$-4 - 8y - 3y = 18$$

$$-11y = 18 + 4 = 22$$

$$y = \frac{22}{-11} = -2$$

Substituting $y = -2$ into equation (1) gives:

$$x + 2(-2) = -1$$

$$x - 4 = -1$$

$$x = -1 + 4 = 3$$

Thus **$x = 3$** and **$y = -2$** is the solution to the simultaneous equations.

(Check: In equation (2), since $x = 3$ and $y = -2$, LHS $= 4(3) - 3(-2)$ $= 12 + 6 = 18 = $ RHS)

Example 5.41

Solve the following equations for x and y by elimination:

$$x + 2y = -1 \qquad (1)$$

$$4x - 3y = 18 \qquad (2)$$

If equation (1) is multiplied throughout by 4 the coefficient of x will be the same as in equation (2), giving:

$$4x + 8y = -4 \ (3)$$

Subtracting equation (3) from equation (2) gives:

$$4x - 3y = 18 \qquad (2)$$

$$4x + 8y = -4 \qquad (3)$$

$$\overline{ 0 - 11y = 22}$$

Hence

$$y = \frac{22}{-11} = -2$$

(Note, in the above subtraction, $18 - -4 = 18 + 4 = 22$.)

Substituting $y = -2$ into either equation (1) or equation (2) will give $x = 3$ as in method (a). The solution $x = 3$, $y = -2$ is the only pair of values that satisfies both of the original equations.

Example 5.42

Solve $\qquad 7x - 2y = 26 \qquad (1)$

and $\qquad 6x + 5y = 29 \qquad (2)$

When equation (1) is multiplied by 5 and equation (2) by 2 the coefficients of y in each equation are numerically the same, i.e. 10, but are of opposite sign.

$5 \times$ equation (1) gives: $\qquad 35x - 10y = 130 \qquad (3)$

$2 \times$ equation (2) gives: $\qquad 12x + 10y = 58 \qquad (4)$

Adding equations (3) and (4) gives: $\overline{\quad 47x + \ 0 \ = 188\quad}$

Hence $x = \dfrac{188}{47} = 4$

Substituting $x = 4$ in equation (1) gives: $(7 \times 4) - 2y = 26$

Hence $\quad 28 - 26 = 2y$ or $y = 1$

Checking, by substituting $x = 4$, $y = 1$ in equation (2), gives:

$$\text{LHS} = 6(4) + 5(1) = 24 + 5 = 29 = \text{RHS}$$

Thus the solution is $x = 4$, $y = 1$, since these values maintain the equality when substituted in both equations.

Example 5.43

Solve

$$\frac{2}{x} + \frac{3}{y} = 7 \qquad (1)$$

and

$$\frac{1}{x} - \frac{4}{y} = -2 \qquad (2)$$

In this type of equation a substitution can initially be made.

Let $\dfrac{1}{x} = a$ and $\dfrac{1}{y} = b$

Thus equation (1) becomes: $\qquad 2a + 3b = 7 \qquad (3)$

and equation (2) becomes: $\qquad a - 4b = -2 \qquad (4)$

Multiplying equation (4) by 2 gives: $\quad 2a - 8b = -4 \qquad (5)$

Subtracting equation (5) from equation (3) gives:

$$0 + 11b = 11$$

i.e. $\qquad\qquad\qquad\qquad b = 1$

Substituting $b = 1$ in equation (3) gives:

$$2a + 3 = 7$$
$$2a = 7 - 3 = 4$$

i.e. $\qquad a = 2$

Checking, substituting $a = 2$, $b = 1$ in equation (4) gives:

$$\text{LHS} = 2 - 4(1) = 2 - 4 = -2 = \text{RHS}$$

Hence $a = 2$, $b = 1$

However, since $\dfrac{1}{x} = a$ then $x = \dfrac{1}{a} = \dfrac{1}{2}$

and since $\dfrac{1}{y} = b$ then $y = \dfrac{1}{b} = \dfrac{1}{1} = 1$

Hence the solution is **$x = \frac{1}{2}$, $y = 1$** which may be checked in the original equations.

Test your knowledge 5.7

Solve the following simultaneous equations

1. $2x + 5y = -7$

   $x + 3y = 4$

2. $3p = 2q$

   $4p + q + 11 = 0$

3. $2.5x + 0.75 - 3y = 0$

   $1.6x = 1.08 - 1.2y$

4. $\dfrac{x-1}{3} + \dfrac{y+2}{5} = \dfrac{2}{15}$

   $\dfrac{1-x}{6} + \dfrac{5+y}{2} = \dfrac{5}{6}$

Problems involving simultaneous equations

Some engineering problems can only be solved using simultaneous equations. In such cases, you will find that you have two equations and two unknowns. You can then decide whether to use *substitution* or *elimination* techniques to solve them.

Example 5.44

The law connecting friction F and load L for an experiment is of the form $F = aL + b$, where a and b are constants. When $F = 5.6$, $L = 8.0$ and when $F = 4.4$, $L = 2.0$. Find the values of a and b and the value of F when $L = 6.5$.

Substituting $F = 5.6$, $L = 8.0$ into $F = aL + b$ gives:

$$5.6 = 8.0a + b \tag{1}$$

Substituting $F = 4.4$, $L = 2.0$ into $F = aL + b$ gives:

$$4.4 = 2.0a + b \tag{2}$$

Subtracting equation (2) from equation (1) gives:

$$1.2 = 6.0\ a \text{ from which } a = 12/60 = 1/5 = 0.2$$

Substituting $a = 1/5 = 0.2$ into equation (1) gives:

$$5.6 = 8.0 \times 0.2 + b \text{ thus } 5.6 = 1.6 + b$$

I.e. $\quad 5.6 - 1.6 = b$ from which $b = 4$

Checking, substituting $a = 0.2$, $b = 4$ in equation (2), gives:

$$\text{RHS} = 2.0 \times 0.2 + 4 = 0.4 + 4 = 4.4 = \text{LHS}$$

Hence **$a = 0.2$** and **$b = 4$**

When $L = 6.5$, $F = aL + b = (0.2 \times 6.5) + 4 = 1.3 + 4$, i.e. $F = \mathbf{5.30}$

Example 5.45

The distance s metres from a fixed point of a vehicle travelling in a straight line with constant acceleration, a m/s^2, is given by $s = ut + \frac{1}{2}at^2$, where u is the initial velocity in m/s and t the time in seconds. Determine the initial velocity and the acceleration given that $s = 42$ m when $t = 2$ s and $s = 144$ m when $t = 4$ s. Find also the distance travelled after 3 s

Substituting $s = 42$, $t = 2$ into $s = ut + \frac{1}{2}at^2$ gives: $42 = 2u + \frac{1}{2}a(2)^2$

i.e. $\quad\quad\quad 42 = 2u + 2a \tag{1}$

Substituting $s = 144$, $t = 4$ into $s = ut + \frac{1}{2}at^2$ gives: $144 = 4u + \frac{1}{2}a(4)^2$

i.e. $\quad\quad\quad 144 = 4u + 8a \tag{2}$

Multiplying equation (1) by 2 gives: $84 = 4u + 4a$ (3)

Subtracting equation (3) from equation (2) gives: $60 = 0 + 4a$

i.e. $\quad a = 60/4 = 15$

Substituting $a = 15$ into equation (1) gives: $42 = 2u + 2(15)$

i.e. $\quad 42 - 30 = 2u$ therefore $u = 12/2 = 6$

Hence the initial velocity, **$u = 6$ m/s** and the acceleration, **$a = 15$ m/s^2**

Distance travelled after 3 s is given by $s = ut + \frac{1}{2}at^2$ where $t = 3$, $u = 6$ and $a = 15$. Hence $s = 6(3) + \frac{1}{2}(15)(3)^2 = 18 + 67.5$

i.e. distance travelled after 3 s **$= 85.5$ m**

Activity 5.2

An electronic equipment manufacturer has asked you to advise on the application of a new type of amplifier. The law connecting output current, i, to input voltage, v, is known to be of the form $i = av + b$, where a and b are constants. Experiments show that when i = 1.5 A, v = 0.02 V and when i = 4 A, v = 0.04 V. The manufacturer has asked you to find the values of the constants a and b and also the value of the input voltage required to produce an output current of exactly 5 A. Provide a fully worked solution.

Evaluation of formulae

The statement $v = u + at$ is said to be a *formula* for v in terms of u, a and t. v, u, a and t are *symbols*. The single term on the left-hand side of the equation, v, is called the *subject of the formula*. Provided values are given for all the symbols in a formula except one, the remaining symbol can be made the subject of the formula and may be evaluated by using a calculator.

Example 5.46

In an electrical circuit the voltage V is given by Ohm's law, i.e. $V = IR$. Find, correct to 4 significant figures, the voltage when I = 5.36 A and R = 14.76 Ω

$$V = IR = (5.36)(14.76)$$

Hence voltage, V = **79.11 V**, correct to 4 significant figures.

Example 5.47

The surface area A of a hollow cone is given by $A = \pi r l$. Determine the surface area when r = 3.0 cm, l = 8.5 cm and π = 3.14.

$$A = \pi r l = (3.14)(3.0)(8.5) \text{ cm}^2$$

Hence surface area, A = **80.07 cm^2**

Example 5.48

Velocity v is given by $v = u + at$. If u = 9.86 m/s, a = 4.25 m/s^2 and t = 6.84 s, find v, correct to 3 significant figures.

$$v = u + at = 9.86 + (4.25)(6.84)$$

$$= 9.86 + 29.07 = 38.93$$

Hence velocity, v = **38.9 m/s**, correct to 3 significant figures.

1   The area, A, of a circle is given by $A = \pi r^2$. Determine the area correct to 2 decimal places, given $\pi = 3.142$ and $r = 5.23$ m.

2   Force F newtons is given by the formula $F = (Gm_1m_2)/d^2$, where m_1 and m_2 are masses, d their distance apart and G is a constant. Find the value of the force given that $G = 6.67 \times 10^{-11}$, $m_1 = 7.36$, $m_2 = 15.5$ and $d = 22.6$. Express the answer in standard form, correct to 3 significant figures.

3   The volume V cm^3 of a right circular cone is given by $V = 1/3\ \pi r^2h$. Given that $r = 4.321$ cm, $h = 18.35$ cm and $\pi = 3.142$, find the volume correct to 4 significant figures.

4   The current I amperes in an a.c. circuit is given by

$$I = \frac{V}{\sqrt{(R^2 + X^2)}}$$

Evaluate the current when $V = 250$ V, $R = 11.0\ \Omega$ and $X = 16.2\ \Omega$.

Example 5.49

The power P watts dissipated in an electrical circuit may be expressed by the formula $P = V^2/R$. Evaluate the power, correct to 3 significant figures, given that $V = 17.48$ V and $R = 36.12\ \Omega$

$$P = \frac{V^2}{R} = \frac{17.48^2}{36.12} = \frac{305.6}{36.12} = 8.46$$

Hence power, $P =$ **8.46 W**, correct to 3 significant figures.

Example 5.50

The time of swing, t seconds, of a simple pendulum is given by

$$t = 2\pi\sqrt{\frac{l}{g}}$$

Determine the time, correct to 3 decimal places, given that $\pi = 3.142$, $l = 12.0$ and $g = 9.81$

$$t = 2\pi\sqrt{\frac{l}{g}} = 2 \times 3.142 \times \sqrt{\frac{12.0}{9.81}}$$

$$= 6.284\sqrt{(1.223)} = 6.950$$

Hence time $t =$ **6.950 seconds**, correct to 3 decimal places.

Problems involving equations

When the symbol other than the subject is required to be calculated it is usual to rearrange the formula to make a new subject. This rearranging process is called *transposing the formula* or simply *transposition*. The rules used for transposition of formulae are the same as those used for the solution of simple equations – basically, *that the equality of an equation must be maintained*.

Example 5.51

Transpose $p = q + r + s$ to make r the subject.

The aim is to obtain r on its own on the left-hand side (LHS) of the equation. Changing the equation around so that r is on the LHS gives:

$$q + r + s = p \tag{1}$$

Subtracting $(q + s)$ from both sides of the equation gives:

$$q + r + s - (q + s) = p - (q + s)$$

Thus $q + r + s - q - s = p - q - s$

i.e. $$r = p - q - s \tag{2}$$

It is important to note that, with simple equations, a quantity can be moved from one side of an equation to the other with an appropriate *change of sign*. This is illustrated in Example 5.51 where equation (2) follows immediately from equation (1) by simply moving values to the other side of the 'equal sign' and changing their sign.

Example 5.52

When a body falls freely through a height h, the velocity v is given by $v^2 = 2gh$. Express this formula with h as the subject.

Rearranging gives: $2gh = v^2$

Dividing both sides by 2g gives: $\dfrac{2gh}{2g} = \dfrac{v^2}{2g}$

i.e. $h = \dfrac{v^2}{2g}$

Example 5.53

A formula for the distance moved by a body is given by $s = \frac{1}{2}(v + u)t$. Rearrange the formula to make u the subject.

Rearranging gives: $\frac{1}{2}(v + u)t = s$

Multiplying both sides by 2 gives: $(v + u)t = 2s$

Dividing both sides by t gives: $\dfrac{2(v+u)t}{t} = \dfrac{2s}{t}$

i.e. $v + u = \dfrac{2s}{t}$

Hence $u = \dfrac{2s}{t} - v$ or $u = \dfrac{2s - vt}{t}$

Test your knowledge 5.9

Make the symbol indicated in brackets the subject of each of the following formulae:

1 Make m the subject of the formula,
$$a = \frac{f}{m}$$

2 Make a the subject of the formula, $R_t = R_o (1 + a\,t)$

3 Make g the subject of the formula,
$$t = 2\pi \sqrt{\frac{l}{g}\frac{f}{m}}$$

4 Make R the subject of the formula,
$$I = \frac{E - e}{R + r}$$

5 Make a the subject of the formula,
$$p = \frac{a^2 x + a^2 y}{r}$$

Example 5.54

The impedance of an a.c. circuit is given by $Z = \sqrt{(R^2 + X^2)}$. Make the reactance, X, the subject.

Rearranging gives: $\sqrt{(R^2 + X^2)} = Z$

Squaring both sides gives: $R^2 + X^2 = Z^2$

Rearranging gives: $X^2 = Z^2 - R^2$

Taking the square root of both sides gives: $X = \sqrt{(Z^2 - R^2)}$

Problems involving equations

An *equation* is a statement that two quantities are equal. To '*solve an equation*' means 'to find the value of the unknown'. The value of the unknown is called the *root* of the equation.

A *quadratic equation* is one in which the highest power of the unknown quantity is 2. For example, $x^2 - 3x + 1 = 0$ is a quadratic equation. There are several methods of *solving quadratic equations* including *factorization* and using the *quadratic formula*.

Factorization

Multiplying out $(2x + 1)(x - 3)$ gives:

$$2x^2 - 6x + x - 3, \quad \text{i.e. } 2x^2 - 5x - 3.$$

The reverse process of moving from $2x^2 - 5x - 3$ to $(2x + 1)(x - 3)$ is called *factorizing*.

If the quadratic expression can be factorized this provides the simplest method of solving a quadratic equation. For example, if $2x^2 - 5x - 3 = 0$, then, by factorizing:

$$(2x + 1)(x - 3) = 0$$

Hence either $(2x + 1) = 0$, i.e. $x = \frac{1}{2}$

or $(x - 3) = 0$, i.e. $x = 3$

Unfortunately the technique of factorizing is often one of 'trial and error' but we can often make an intelligent guess as to the factors and this can often speed up the process of arriving at a solution.

Example 5.55

Solve the equation $x^2 + 2x - 8 = 0$ by factorization.

$x^2 + 2x - 8 = 0$. The factors of x^2 are x and x. These are placed in two brackets thus:

$$(x \ldots\ldots)(x \ldots\ldots)$$

The factors of -8 are $+8$ and -1, or -8 and $+1$, or $+4$ and -2, or -4 and $+2$. The only combination to give a middle term of $+2x$ is $+4$ and -2, i.e.

$$x^2 + 2x - 8 = (x - 2)(x + 4)$$

(Note that the product of the two inner terms added to the product of the two outer terms must equal the middle term, $+2x$ in this case.)

The quadratic equation $x^2 + 2x - 8 = 0$ thus becomes $(x + 4)(x - 2) = 0$.

Since the only way that this can be true is for either the first or the second, or both factors to be zero, then either

$$(x + 4) = 0 \quad \text{i.e. } x = -4$$

or $(x - 2) = 0$ i.e. $x = 2$

Hence the roots of $x^2 + 2x - 8 = 0$ are $x = **-4**$ and $x = **2**$.

Example 5.56

Solve the equation $3x^2 - 11x - 4 = 0$ by factorization.

The factors of $3x^2$ are $3x$ and x. These are placed in brackets thus:

$$(x \ldots\ldots)(3x \ldots\ldots)$$

The factors of -4 are -4 and $+1$, or $+4$ and -1, or -2 and 2.

Remembering that the product of the two inner terms added to the product of the two outer terms must equal $-11x$, the only combination to give this is -4 and $+1$, i.e.

$$3x^2 - 11x - 4 = (3x + 1)(x - 4)$$

The quadratic equation $3x^2 - 11x - 4 = 0$ thus becomes $(3x + 1)(x - 4) = 0$.

Hence,   either $(3x + 1) = 0$ i.e.

  or $(x - 4) = 0$ i.e. ***x* = 4**

  and both solutions may be checked in the original equation.

Example 5.57

Determine the roots of $x^2 - 6x + 9 = 0$ by factorization.

$$x^2 - 6x + 9 = 0 \quad \text{hence} \quad (x - 3)(x - 3) = 0, \quad \text{i.e. } (x - 3)^2 = 0$$

The left-hand side of this expression is known as *a perfect square*. Hence ***x* = 3** is the *only* root of the equation $x^2 - 6x + 9 = 0$.

Example 5.58

Determine the roots of $4x^2 - 25 = 0$ by factorization.

The left-hand side of the equation, $4x^2 - 25$, is *the difference of two squares*, $(2x)^2$ and $(5)^2$. Hence

$(2x + 5)(2x - 5) = 0$

Thus either $(2x + 5) = 0$ i.e. $x = -\dfrac{5}{2}$

or $(2x - 5) = 0$ i.e. $x = \dfrac{5}{2}$

Example 5.59

Determine the roots of $100x^2 - 256 = 0$ by factorization.

$$100x^2 - 256x = (10x - 16)(10 - 16) = 0, \quad \text{i.e. } (10x - 16)^2 = 0$$

Once again, the left-hand side of this expression is known as *a perfect square*.

$x = 16/10$ is the *only* root of the equation thus ***x* = 1.6**

Solution of quadratic equations by formulae

Let the general form of a quadratic equation be given by:

$$ax^2 + bx + c = 0$$

where a, b and c are constants.

Dividing $ax^2 + bx + c = 0$ by a gives:

$$x^2 + \frac{b}{a}x + \frac{c}{a} = 0$$

Rearranging gives:

$$\left(x^2 + \frac{b}{a}x\right) + \frac{c}{a} = 0$$

Adding to each side of the equation the square of half the coefficient of the term in x to make the LHS a perfect square gives:

$$\left(x^2 + \frac{b}{a}x\right) + \left(\frac{b}{2a}\right)^2 + \frac{c}{a} = \left(\frac{b}{2a}\right)^2$$

Rearranging gives:

$$\left(x^2 + \frac{b}{a}x + \left(\frac{b}{2a}\right)^2\right) = \left(\frac{b}{2a}\right)^2 - \frac{c}{a}$$

$$\left(x + \frac{b}{2a}\right)^2 = \left(\frac{b}{2a}\right)^2 - \frac{c}{a} = \frac{b^2 - 4ac}{4a^2}$$

$$\left(x + \frac{b}{2a}\right)^2 = \frac{b^2 - 4ac}{4a^2}$$

Taking square roots of both sides of the equation gives:

$$x + \frac{b}{2a} = \sqrt{\frac{b^2 - 4ac}{4a^2}}$$

$$x + \frac{b}{2a} = \pm\frac{\sqrt{b^2 - 4ac}}{2a}$$

$$x = -\frac{b}{2a} \pm \frac{\sqrt{b^2 - 4ac}}{2a}$$

$$x = \frac{-b \pm \sqrt{b^2 - 4ac}}{2a}$$

This is known as the *quadratic formula* and the roots are given by:

$$x = \frac{-b + \sqrt{b^2 - 4ac}}{2a} \quad \text{and} \quad x = \frac{-b - \sqrt{b^2 - 4ac}}{2a}$$

Solve the following equations by using the quadratic formula:

1 $2x^2 + 5x - 3 = 0$

2 $3x^2 - 11x - 4 = 0$

3 $x^2 - 27x + 38 = 0$ (correct to four significant figures)

Example 5.60

Solve $x^2 + 2x - 8 = 0$ by using the quadratic formula.

Comparing $x^2 + 2x - 8 = 0$ with $ax^2 + bx + c = 0$ gives $a = 1$, $b = 2$ and $c = -8$. Substituting these values into the quadratic formula

$$x = \frac{-b \pm \sqrt{b^2 - 4ac}}{2a}$$

gives:

$$x = \frac{-2 \pm \sqrt{(2)^2 - 4(1)(-8)}}{2(1)}$$

$$x = \frac{-2 \pm \sqrt{36}}{2}$$

Hence $x = 4/2 = $ **2** or $-8/2 = $ **-4**

Problems involving equations

Example 5.61

Calculate the diameter of a solid cylinder that has a height of 82.0 cm and a total surface area of 2.0 m²

Total surface area of a cylinder = curved surface area + 2 circular ends

$$= 2\pi rh + 2\pi r^2 \text{ (where } r = \text{radius and } h = \text{height)}$$

Since the total surface area = 2.0 m² and the height $h = 82$ cm or 0.82 m, then $2.0 = 2\pi r(0.82) + 2\pi r^2$

i.e. $2\pi r^2 + 2\pi r(0.82) - 2.0 = 0$

Dividing throughout by 2π gives:

$$r^2 + 0.82r - 1/\pi = 0$$

Using the quadratic formula: $x = \dfrac{-b \pm \sqrt{b^2 - 4ac}}{2a}$

gives: $r = \dfrac{-0.82 \pm \sqrt{(0.82)^2 - 4(1)(-1/\pi)}}{2(1)}$

$$r = \frac{-0.82 \pm \sqrt{1.9456}}{2}$$

$$r = \frac{-0.82 \pm 1.3948}{2} = \frac{-0.82 + 1.3948}{2} \text{ or } \frac{-0.82 - 1.3948}{2}$$

Hence $r = 0.2874$ or -1.1074. We can ignore the negative result and use the positive solution to determine the diameter of the cylinder. Thus:

$d = 2 \times 0.2874 = 0.5748$ m = **57.5 cm** correct to three significant figures.

Example 5.62

A shed is 4.0 m long and 2.0 m wide. A concrete path of constant width is laid all the way around the shed. If the area of the path is 9.50 m² calculate its width to the nearest centimetre.

Figure 5.1 shows a plan view of the shed with its surrounding path of width t metres.

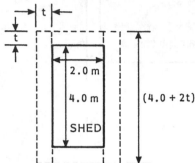

Figure 5.1

Area of path = $2(2.0 \times t) + 2t(4.0 + 2t)$ i.e. $9.50 = 4.0t + 8.0t + 4t^2$

or $4t^2 + 12.0t - 9.50 = 0$

Using the quadratic formula: $x = \dfrac{-b \pm \sqrt{b^2 - 4ac}}{2a}$

gives:

$$t = \dfrac{-12 \pm \sqrt{(12)^2 - 4(4)(-9.5)}}{2(4)}$$

$$t = \dfrac{-12 \pm \sqrt{296}}{8}$$

$$t = \dfrac{-12 \pm 17.20465}{2} = \dfrac{-12 + 17.20465}{2} \text{ or } \dfrac{-12 - 17.20465}{2}$$

Hence $t = 0.6506$ or -3.65058

We can ignore the negative result and use the positive solution to determine the width of the path:

$t = 0.651$ m = **65 cm** correct to the nearest centimetre.

1 The area of a rectangle is 23.6 cm² and its width is 3.10 cm shorter than its length. Determine the dimensions of the rectangle, correct to three significant figures.

2 The bending moment M of a beam at a point in a beam is given by:

$$M = \dfrac{3x(20-x)}{2}$$

where x metres is the distance from the point of support. Determine the values of x when the bending moment is 50 Nm.

3 If the total surface area of a solid cone is 486.2 cm² and its slant height is 15.3 cm, determine its base diameter.

Activity 5.3

You have been asked to advise on the construction of a vertical cylindrical storage tank that is to be fitted in a chemical plant. The storage tank is to be located in a room which has a height of 2.5 m. Determine the diameter of the tank if it is to have a total surface area of 15 m². Prepare a report for the engineering manager and include a fully worked solution to the problem. (You can ignore the thickness of the tank in your calculations.)

Trigonometry

Trigonometry is the branch of mathematics that deals with the measurement of sides and angles of triangles, and their relationships with each other.

Trigonometric ratios

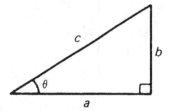

Figure 5.2

With reference to the *right-angled triangle* shown in Figure 5.2:

(i) $\sin\theta = \dfrac{\text{opposite}}{\text{hypotenuse}}$, i.e. $\sin\theta = \dfrac{b}{c}$

(ii) $\cos\theta = \dfrac{\text{adjacent}}{\text{hypotenuse}}$, i.e. $\cos\theta = \dfrac{a}{c}$

(iii) $\tan\theta = \dfrac{\text{opposite}}{\text{adjacent}}$, i.e. $\tan\theta = \dfrac{b}{a}$

From above,

$$\frac{\sin\theta}{\cos\theta} = \frac{b/c}{a/c} = \frac{b}{a}, \text{ i.e. } \tan\theta = \frac{\sin\theta}{\cos\theta}$$

Example 5.63

From Figure 5.3, find $\sin D$, $\cos D$ and $\tan X$.

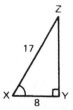

Figure 5.3

By Pythagoras' theorem, $17^2 = 8^2 + YZ^2$, from which, $YZ = \sqrt{(17^2 - 8^2)} = 15$

$$\sin X = \frac{YZ}{XZ} = \frac{15}{17} \text{ or } \mathbf{0.8824}$$

$$\cos X = \frac{XY}{XZ} = \frac{8}{17} \text{ or } \mathbf{0.4706}$$

$$\tan X = \frac{YZ}{XY} = \frac{15}{8} \text{ or } \mathbf{1.875}$$

Example 5.64

If cos X in Figure 5.4 is 9/41 determine the value of the other two trigonometric ratios.

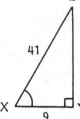

Figure 5.4

Figure 5.4 shows a right-angled triangle. Since cos X = 9/41, then XY = 9 units and XZ = 41 units.

Using Pythagoras' theorem to find the remaining side gives: $41^2 = 9^2 + YZ^2$ from which YZ = $\sqrt{(41^2 - 9^2)}$ = 40 units

Thus

$$\sin X = \frac{YZ}{XZ} = \frac{40}{41} \text{ or } \mathbf{0.9756}$$

$$\tan X = \frac{YZ}{XY} = \frac{40}{9} \text{ or } \mathbf{4.444}$$

Example 5.65

If sin θ = 0.625 and cos θ = 0.500 determine, without using a calculator, the value of tan θ.

$$\tan \theta = \frac{\sin \theta}{\cos \theta} = \frac{0.625}{0.5} = \frac{5/8}{1/2} = \frac{5 \times 2}{1 \times 8} = \frac{10}{8} = 1.25$$

Evaluating trigonometric ratios

The values of the sine, cosine, and tangent of an angle can easily be found using a calculator. The following values, correct to 4 decimal places, may be checked:

$$\sin 18° = 0.3090 \qquad \cos 56° = 0.5592$$

$$\tan 29° = 0.5543 \qquad \sin 172° = 0.1392$$

$$\cos 115° = -0.4226 \qquad \tan 178° = -0.0349$$

$$\sin 241.63° = -0.8799 \qquad \cos 331.78° = 0.8811$$

$$\tan 296.42° = -2.0127$$

Sometimes angles are expressed in degrees and *minutes* rather than decimal parts of a degree. To evaluate, say, sine 42°23' using a calculator means finding:

	$\sin\left(23\dfrac{23}{60}\right)^{\circ}$

Test your knowledge 5.12

Evaluate, correct to four decimal places:

(a) sin 154.21°

(b) cos 211.46°

(c) 4.2 tan 49°26'−3.7 sin 66°11'

$\sin\left(23\dfrac{23}{60}\right)^{\circ}$

since there are 60 minutes in 1 degree.

23/60 = 0.3833, thus 42°23' = 42.3833°

Thus

sin 42°23' = sin 42.383 = 0.6741 (correct to four decimal places).

Problems involving right angled triangles

Right angled triangles are found in many engineering applications. To 'solve a right-angled triangle' means 'to find the unknown sides and angles'. This is achieved by using the theorem of Pythagoras, and/or trigonometric ratios (sine, cosine and tangent).

Example 5.66

An electricity pylon stands on horizontal ground. At a point 80 m from the base of the pylon, the angle of elevation of the top of the pylon is 23°. Calculate the height of the pylon to the nearest metre.

Figure 5.5 shows the pylon AB and the angle of elevation of A from point C is 23°.

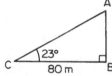

Figure 5.5

Hence height of pylon AB = 80 tan 23°

$$\tan 23^{\circ} = \frac{AB}{BC} = \frac{AB}{80} = 80(0.4245) = 33.96 \text{ m}$$

$$= \textbf{34 m} \text{ to the nearest metre.}$$

Test your knowledge 5.13

1 If the angle of elevation of the top of a vertical 30 m high aerial is 32°, how far is it to the aerial?

2 A surveyor measures the angle of elevation of the top of a perpendicular building as 19°. He moves 120 m nearer the building and finds the angle of elevation is now 47°. Determine the height of the building.

Activity 5.4

A 120 m radio mast is to be located on flat ground close to a transmitter site. The mast is to be supported by four guy wires that run from the top of the mast to concrete posts that are each 25 m from the base of the mast. Determine the angle that each guy wire makes with the ground and the total length of guy wire that is required to complete the installation. Also estimate the % saving in guy wire if the guy wires are anchored to the mast at a point that is 30 m below the top. Present your answer in the form of a fully worked solution and include labelled sketches.

Graphs of trigonometric functions

By drawing up tables of values from $0°$ to $360°$, graphs of $y = \sin A$, $y = \cos A$ and $y = \tan A$ may be plotted. Values obtained with a calculator (correct to three decimal places – which is more than sufficient for plotting graphs), using $30°$ intervals, are shown below, with the respective graphs shown in Figure 5.6.

(a) $y = \sin A$

A	$0°$	$30°$	$60°$	$90°$	$120°$	$150°$	$180°$
$\sin A$	0	0.500	0.866	1	0.866	0.500	0

A	$210°$	$240°$	$270°$	$300°$	$330°$	$360°$
$\sin A$	−0.500	−0.866	−1	−0.866	−0.500	0

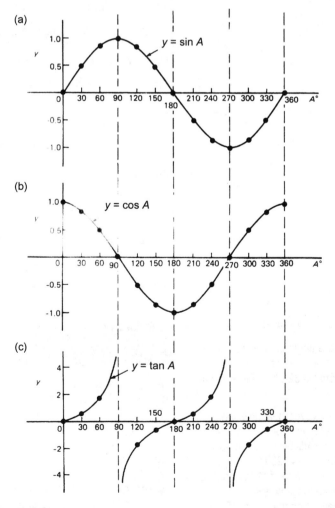

Figure 5.6

(b) $y = \cos A$

A	0°	30°	60°	90°	120°	150°	180°
sin A	1.000	0.866	0.500	0	−0.500	−0.866	−1

A	210°	240°	270°	300°	330°	360°
sin A	−0.866	−0.500	0	0.500	0.866	1

(c) $y = \tan A$

A	0°	30°	60°	90°	120°	150°	180°
sin A	0	0.577	1.732	∞	−1.732	−0.577	0

A	210°	240°	270°	300°	330°	360°
sin A	0.577	1.732	∞	−1.732	−0.577	0

If you take a careful look at Figure 5.6 you should notice that:

(i)   Sine and cosine graphs oscillate between peak values of ±1

(ii) The cosine curve is the *same shape* as the sine curve but is displaced by 90°

(iii) The sine and cosine curves are continuous and they *repeat* at intervals of 360°; the tangent curve appears to be discontinuous and repeats at intervals of 180°.

Angles of any magnitude

(i)   Figure 5.7 shows rectangular axes XX' and YY' intersecting at origin 0. As with graphical work, measurements made to the right and above 0 are positive while those to the left and downwards are negative. Let OA be free to rotate about 0. By convention, when OA moves anticlockwise angular measurement is considered positive, and vice-versa.

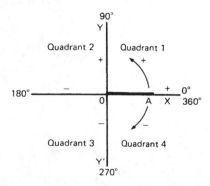

Figure 5.7

(ii) Let OA be rotated anticlockwise so that θ_1 is any angle in the first quadrant and let perpendicular AB be constructed to form the right-angled triangle OAB (see Figure 5.8). Since all three sides of the triangle are positive, all six trigonometric ratios are positive in the first quadrant. (Note: OA is always positive since it is the radius of a circle.)

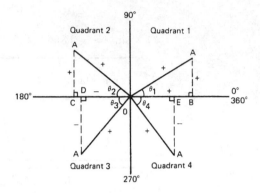

Figure 5.8

(iii) Let OA be further rotated so that θ_2 is any angle in the second quadrant and let AC be constructed to form the right-angled triangle OAC. Then:

$$\sin \theta_2 = +/+ = +, \ \cos \theta_2 = -/+ = -, \ \tan \theta_2 = +/- = -$$

(iv) Let OA be further rotated so that θ_3 is any angle in the third quadrant and let AD be constructed to form the right-angled triangle OAD. Then:

$$\sin \theta_3 = -/+ = -, \ \cos \theta_3 = -/+, = -, \ \tan \theta_3 = -/- = +$$

(v) Let OA be further rotated so that θ_4 is any angle in the fourth quadrant and let AE be constructed to form the right-angled triangle OAE. Then:

$$\sin \theta_4 = -/+ = -, \ \cos \theta_4 = +/+ = +, \ \tan \theta_4 = -/+ = -$$

(vi) The results obtained in (ii) to (v) are summarized in Figure 5.9. The letters underlined spell the word CAST when starting in the fourth quadrant and moving in an anticlockwise direction.

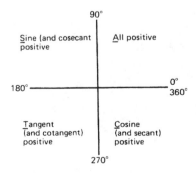

Figure 5.9

(vii) In the first quadrant of Figure 5.7 all the curves have positive values; in the second only sine is positive; in the third only tangent is positive; in the fourth only cosine is positive (exactly as summarized in Figure 5.9).

A knowledge of angles of any magnitude is needed when finding, for example, all the angles between $0°$ and $360°$ whose sine is, say, 0.3261. If 0.3261 is entered into a calculator and then the inverse sine key pressed (or $\sin^{-1}$ key) the answer $19.03°$ appears. However there is a second angle between $0°$ and $360°$ that the calculator does not give. Sine is also positive in the second quadrant (either from CAST or from Figure 5.6(a)).

The other angle is shown in Figure 5.10 as angle θ where $\theta = 180° - 19.03° = 160.97°$. Thus $19.03°$ *and* $160.97°$ are the angles between $0°$ and $360°$ whose sine is 0.3261 (check that $\sin 160.97° = 0.3261$ on your calculator).

Be careful! Your calculator only gives you one of these answers. The second answer needs to be deduced from a knowledge of angles of any magnitude, as shown in the following problem and in Figure 5.11.

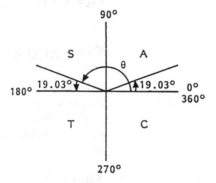

Figure 5.10

Example 5.67

Determine all the angles between $0°$ and $360°$ (a) whose sine is -0.4638 and (b) whose tangent is 1.7629.

(a) The angles whose sine is 0.4638 occurs in the third and fourth quadrants since sine is negative in these quadrants (see Figure 5.11 (a)). From Figure 5.11(b), θ = arcsin 0.4638 = $27°38'$. (Note that arcsin 0.4638 simply means *the angle whose sine is* 0.4638.) Measured from $0°$, the two angles between $0°$ and $360°$ whose sine is -0.4638 are $180° + 27°38'$, i.e. **207°38'** and $360° - 27°38'$, i.e. **332°22'**. (Note that a calculator generally only gives one answer, i.e. $-27.632588°$.)

(b) A tangent is positive in the first and third quadrants (see Figure 5.11 (c)). From Figure 5.11(d), θ = arctan 1.7629 = $60°26'$. Measured from $0°$, the two angles between $0°$ and $360°$ whose tangent is 1.7629 are **60°26'** and $180° + 60°26'$, i.e., **240°26'**.

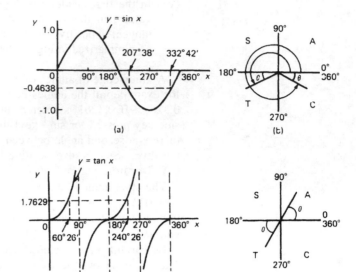

Figure 5.11

Sine and cosine rules

If a triangle is *right angled*, trigonometric ratios and the theorem of Pythagoras may be used for its solution. However, for a *non right-angled triangle*, trigonometric ratios and Pythagoras' theorem *cannot* be used. Instead, two rules, called the sine rule and the cosine rule, are used.

Sine rule

With reference to triangle ABC of Figure 5.12, the *sine rule* states:

$$\frac{a}{\sin A} = \frac{b}{\sin B} = \frac{c}{\sin C}$$

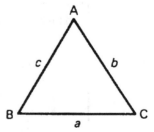

Figure 5.12

The rule may be used only when:

(i)   one side and any two angles are initially given, or

(ii) two sides and an angle (not the included angle) are initially given.

Cosine rule

With reference to triangle ABC of Figure 5.12, the *cosine rule* states:

$$a^2 = b^2 + c^2 - 2bc \cos A$$

$$\text{or } b^2 = a^2 + c^2 - 2ac \cos B$$

$$\text{or } c^2 = a^2 + b^2 - 2ab \cos C$$

The rule may be used only when:

(i) two sides and the included angle are initially given, or

(ii) three sides are initially given.

Area of any triangle

The *area of any triangle* such as ABC of Figure 5.12 is given by:

(i) $\frac{1}{2} \times$ base $\times$ perpendicular height, or

(ii) $\frac{1}{2} ab \sin C$, or $\frac{1}{2} ac \sin B$, or $\frac{1}{2} bc \sin A$, or

(iii) $\sqrt{[s(s - a)(s - b)(s - c)]}$, where $s = \dfrac{a + b + c}{2}$

The method that you use to find the area of any given triangle depends upon the information that you have been given. For example, it's always worth looking to see if you can easily find the perpendicular height – this is easy if you are dealing with a right-angled triangle.

Example 5.68

Determine the area of the triangle shown in Figure 5.13.

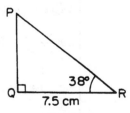

Figure 5.13

Now area = $\frac{1}{2} \times$ base $\times$ perpendicular height

$= \frac{1}{2} \times 7.5 \times$ PQ

But tan 38° = PQ/QR   thus   PQ = QR $\times$ tan 38°

Hence PQ = 7.5 $\times$ tan 38° = 5.86

Thus area = $\frac{1}{2} \times 7.5 \times 5.86$ = **21.975 cm²**

Example 5.69

Solve the triangle PQR and find its area given that QR = 36.5 mm, PR = 29.6 mm and Q = 36°.

Triangle PQR is shown in Figure 5.14.

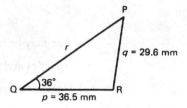

Figure 5.14

Applying the sine rule: $\dfrac{29.6}{\sin 36°} = \dfrac{36.5}{\sin P}$

from which,

$$\sin P = \frac{36.5 \sin 36°}{29.6} = 0.7248$$

Hence P = arcsin 0.7248 = 46°27' or 133°33'

When P = 46°27' and Q = 36° then R = 180° − 46°27' − 36° = 97°33'

When P = 133°33' and Q = 36° then R = 180° − 133°33' − 36° = 10°27'

Thus, in this problem, there are *two* separate sets of results and both are feasible solutions. Such a situation is called the *ambiguous case*.

Case 1. P = 46°27', Q = 36°, R = 97°33', p = 36.5 mm and q = 29.6 mm

From the sine rule: $\dfrac{r}{\sin 97°33'} = \dfrac{29.6}{\sin 36°}$

from which,

$$r = \frac{29.6 \sin 97°33'}{\sin 36°} = \textbf{49.92 mm}$$

Area = ½ pq sin R = ½ (36.5)(29.6) sin 97°33' = **535.5 mm²**

Case 2. P = 133°33', Q = 36°, R = 10°27', p = 36.5 mm and q = 29.6 mm

From the sine rule: $\dfrac{r}{\sin 10°27'} = \dfrac{29.6}{\sin 36°}$

from which,

$$r = \frac{29.6 \sin 10°27'}{\sin 36°} = \textbf{9.134 mm}$$

Area = ½ pq sin R = ½ (36.5)(29.6) sin 10°27' = **97.98 mm²**

Triangle PQR for case 2 is shown in Figure 5.15.

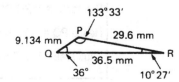

Figure 5.15

1 Solve the triangle ABC and find its area given A = 31°, C = 72° and a = 9.6 mm.

2 Solve triangle DEF and find its area given that EF = 35.0 mm, DE = 25.0 mm and E = 64°

3 Solve triangle XYZ and find its area given x = 12.0 cm, y = 9.0 cm and z = 10.0 cm.

Example 5.70

A triangle ABC has sides a = 9.0 cm, b = 7.5 cm and c = 6.5 cm. Determine its three angles and its area.

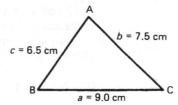

Figure 5.16

Triangle ABC is shown in Figure 5.16. It is usual first to calculate the largest angle to determine whether the triangle is acute or obtuse. In this case the largest angle is *A* (i.e. opposite the longest side).

Applying the cosine rule:

$$a^2 = b^2 + c^2 - 2bc \cos A$$

from which,

$$2bc \cos A = b^2 + c^2 - a^2$$

and

$$\cos A = \frac{b^2 + c^2 - a^2}{2bc} = \frac{7.5^2 + 6.5^2 - 9.0^2}{2 \times 7.5 \times 6.5} = 0.1795$$

Hence *A* = arccos 0.1795 = **79°40'** (or 280°20'), which is obviously impossible). The triangle is thus acute angled since cos *A* is positive.
(If cos *A* had been negative, angle *A* would be obtuse, i.e. lie between 90° and 180°.)

Applying the sine rule: $\dfrac{9.0}{\sin 79°40'} = \dfrac{7.5}{\sin B}$

from which, $\sin B = \dfrac{7.5 \sin 79°40'}{9.0} = \mathbf{0.8198}$

Hence *B* = arcsin 0.8198 = **55°4'**

  C = 180° - 79°40' - 55°4' = **45°16'**

The area can now be calculated from:

$$\text{Area} = \sqrt{[s(s - a)(s - b)(s - c)]}, \text{ where:}$$

$$s = \frac{a + b + c}{2} = \frac{9.0 + 7.5 + 6.5}{2} = 11.5 \text{ cm}$$

Hence area $= \sqrt{[11.5(11.5 - 9.0)(11.5 - 7.5)(11.5 - 6.5)]}$

  $= \sqrt{[11.5(2.5)(4.0)(5.0)]}$

  = **23.98 cm²**

Alternatively, area = ½ *ab* sin *C* = ½ (9.0)(7.5) sin 45°16' = **23.98 cm²**

Problems involving trigonometry

Many practical engineering problems involve the use of trigonometry. When solving these problems you need to be able to recall and use the basic formulae for sine, cosine and tangent as well as the sine and cosine rules.

In some situations, *radians* are used to specify angles rather than degrees.

(i) To convert from radians to degrees multiply by $\dfrac{360}{2\pi} = 57.29$

(ii) To convert from degrees to radians multiply by $\dfrac{2\pi}{360} = 0.0175$

Example 5.71

Two voltage phasors are shown in Figure 5.17. If $V_1 = 40$ V and $V_2 = 100$ V determine the value of their resultant (i.e. length OA) and the angle the resultant makes with V_1.

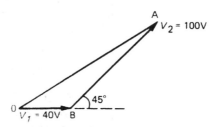

Figure 5.17

Angle   OBA = 180° – 45° = 135°.

Applying the cosine rule:

$$OA^2 = V_1^2 + V_2^2 - 2V_1V_2 \cos OBA$$

$$= 40^2 + 100^2 - \{2(40)(100) \cos 135°\}$$

$$= 1600 + 10000 - \{-5657\}$$

$$= 1600 + 10000 + 5657 = 17257$$

The resultant OA = √(17257) = 131.4 V

Applying the sine rule:

$$\frac{131.1}{\sin 135°} = \frac{100}{\sin AOB}$$

from which,

$$\sin AOB = \frac{100 \sin 135°}{131.4} = 0.5381$$

Hence angle AOB = arcsin 0.5381 = 32°33'

(or 147°27' which is impossible in this case).

Hence the resultant voltage is **131.4 volts** at **32°33'** to V_1.

Example 5.72

In Figure 5.18, PR represents the inclined jib of a crane and is 10.0 m long. PQ is 4.0 m long. Determine the length of tie QR and the inclination of the jib to the vertical.

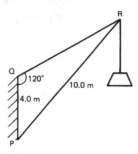

Figure 5.18

Applying the sine rule:

$$\frac{PR}{\sin 120°} = \frac{PQ}{\sin R}$$

from which,

$$\sin R = \frac{PQ \sin 120°}{PR} = \frac{4 \times \sin 120°}{10} = 0.3464$$

Hence R = arcsin 0.3464 = 20°16' (or 159°44', which is impossible in this case).

P = 180° − 120° − 20°16' = **39°44'**, which is the inclination of the jib to the vertical.

Applying the sine rule:

$$\frac{10}{\sin 120°} = \frac{QR}{\sin 39°44'}$$

from which,

$$QR = \frac{10 \sin 39°44'}{\sin 120°} = \textbf{7.38 m}$$

Test your knowledge 5.16

1 A room 8.0 m wide has a span roof which slopes at 33° on one side and 40° on the other. Find the length of the roof slopes, correct to the nearest centimetre.

2 PQ and QR are the phasors representing the alternating currents in two branches of a circuit. Phasor PQ is 20.0 A and is horizontal. Phasor QR (which is joined to the end of PQ to form triangle PQR) is 14.0 A and is at an angle of 35° to the horizontal. Determine the resultant phasor PR and the angle it makes with phasor PQ.

3 A man leaves a point walking at 6.5 km/h in a direction E 20° N (i.e. a bearing of 70°). A cyclist leaves the same point at the same time in a direction E 40° S (i.e. a bearing of 130°) travelling at a constant speed. Find the average speed of the cyclist if the walker and cyclist are 80 km apart after 5 hours.

Example 5.73

The area of a field is in the form of a quadrilateral ABCD as shown in Figure 5.19. Determine its area.

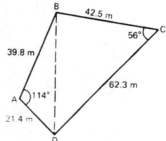

Figure 5.19

A diagonal drawn from B to D divides the quadrilateral into two triangles.

Area of quadrilateral ABCD = area of triangle ABD + area of triangle BCD

$$= \tfrac{1}{2}(39.8)(21.4) \sin 114° + \tfrac{1}{2}(42.5)(62.3) \sin 56°$$

$$= 389.04 + 1097.5 = \textbf{1487 m}^2$$

Graphs

Coordinates

A *graph* is a pictorial representation of information showing how one quantity varies with another related quantity. The most common method of showing a relationship between two sets of data is to use *Cartesian* or *rectangular axes* as shown in Figure 5.20.

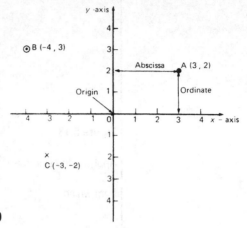

Figure 5.20

The points on a graph are called *coordinates*. Point A in Figure 5.20 has the coordinates (3, 2), i.e. 3 units in the *x* direction and 2 units in the *y* direction. Similarly, point B has coordinates (–4, 3) and C has coordinates (–3, –2). The origin has coordinates (0, 0).

The straight line graph

Let a relationship between two variables *x* and *y* be $y = 3x + 2$.

When $x = 0$, $y = (3 \times 0) + 2 = 2$.

When $x = 1$, $y = (3 \times 1) + 2 = 5$.

When $x = 2$, $y = (3 \times 2) + 2 = 8$, and so on.

Thus coordinates (0, 2), (1, 5) and (2, 8) have been produced from the equation by selecting arbitrary values of *x*, and are shown plotted in Figure 5.21. When the points are joined together a *straight-line graph* results.

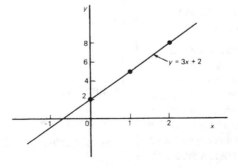

Figure 5.21

The *gradient* or *slope* of a straight line is the ratio of the change in the value of y to the change in the value of x between any two points on the line. If, as x increases ($\rightarrow$), y also increases ($\uparrow$), then the gradient is positive.

In Figure 5.22(a) the gradient of

$$AC = \frac{\text{change in } y}{\text{change in } x} = \frac{CB}{BA} = \frac{7-3}{3-1} = \frac{4}{2} = 2.$$

If as x increases ($\rightarrow$), y decreases ($\downarrow$), then the gradient is negative.

In Figure 5.22(b), the gradient of

$$DF = \frac{\text{change in } y}{\text{change in } x} = \frac{FE}{ED} = \frac{11-3}{-3-0} = \frac{9}{3} = 3.$$

Figure 5.22(c) shows a straight line graph $y = 3$. Since the straight line is horizontal the gradient is zero.

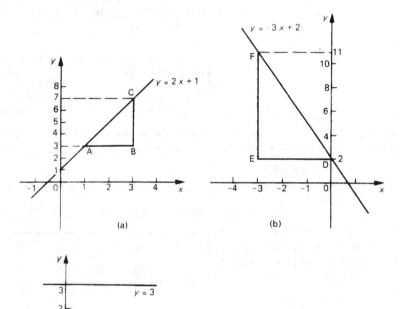

Figure 5.22

The value of y when $x = 0$ is called the *y-axis intercept*. In Figure 5.22(a) the y-axis intercept is 1 and in Figure 5.22(b) is 2.

If the equation of a graph is of the form $y = mx + c$, where m and c are constants, the graph will always be a straight line, m representing the gradient and c the y-axis intercept.

Thus $y = 5x + 2$ represents a straight line of gradient 5 and y-axis intercept 2.

Similarly, $y = -3x - 4$ represents a straight line of gradient -3 and y-axis intercept -4.

General rules to be applied when drawing graphs

(i)   Use a title for the graph that clearly explains what is being illustrated.

(ii) Choose scales such that the graph occupies as much space as possible on the graph paper being used.

(iii) Choose scales so that interpolation is made as easy as possible. Usually scales such as 1 cm = 1 unit, or 1 cm = 2 units, or 1 cm = 10 units are used. Awkward scales such as 1 cm = 3 units or 1 cm = 7 units should not be used.

(iv) The scales need not start at zero, particularly when starting at zero produces an accumulation of points within a small area of the graph paper.

(v)   The coordinates, or points, should be clearly marked. This may be done either by a cross, or a dot and circle, or just by a dot (see Figure 5.20).

(vi) A statement should be made next to each axis explaining the numbers represented with their appropriate units.

(vii) Sufficient numbers should be written next to each axis without cramping.

Example 5.74

Plot the graph $y = 4x + 3$ in the range $x = -3$ to $x = +4$. From the graph, find (a) the value of y when $x = 2.2$, and (b) the value of x when $y = -3$.

Whenever an equation is given and a graph is required, a table giving corresponding values of the variable is necessary. The table is produced as follows:

When $x = -3$, $y = 4x + 3 = (4 \times -3) + 3 = -12 + 3 = -9$

When $x = -2$, $y = (4 \times -2) + 3 = -8 + 3 = -5$, and so on.

x	−3	−2	−1	0	1	2	3	4
y	−9	−5	−1	3	7	11	15	9

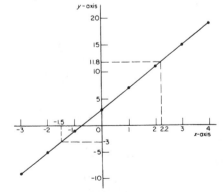

Figure 5.23

The coordinates (−3, −9), (−2, −5), (−1, −1), and so on, are plotted and joined together to produce the straight line shown in Figure 5.23. (Note that the scales used on the x and y axes do not have to be the same.) From the graph:

(a) when $x = 2.2$, $y = \mathbf{11.8}$, and (b) when $y = -3$, $x = \mathbf{-1.5}$

---

1   The equation of a line is $4y = 2x + 5$. Produce a table showing corresponding values of x and y over the range $x = -4$ to $x = 4$. table and plot a graph of y against x. Find the gradient of the graph.

2   Without plotting graphs, determine the gradient and y-axis intercept values of the following equations:

   (a) $y = 7x - 3$

   (b) $3y = -6x + 2$

   (c) $y - 2 = 4x + 9$

   (d) $\dfrac{y}{3} = \dfrac{x}{3} - \dfrac{1}{5}$

   (e) $2x + 9y + 1 = 0$

3   Determine the gradient of the straight line graph passing through the coordinates (−2, 3) and (−1, 3).

Practical problems involving graphs

When a set of coordinate values are given or are obtained experimentally and it is believed that they follow a law of the form: $y = mx + c$, then if a straight line can be drawn reasonably close to most of the coordinate values when plotted, this verifies that a law of the form $y = mx + c$ exists. From the graph, constants m (i.e. gradient) and c (i.e. y-axis intercept) can be determined. This useful technique is called *determination of law*.

Example 5.75

The temperature in degrees Celsius and the corresponding values in degrees Fahrenheit are shown in the table below. Plot this data in the form of a graph. From the graph find (a) the temperature in degrees Fahrenheit at 55°C, (b) the temperature in degrees Celsius at 167°F, (c) the Fahrenheit temperature at 0°C, and (d) the Celsius temperature at 230°F

°C	10	20	40	60	80	100
°F	50	68	104	140	176	212

The coordinates (10, 50), (20, 68), (40, 104), and so on are plotted as shown in Figure 5.40. When the coordinates are joined, a straight line is produced. Since a straight line results there is a *linear* relationship between degrees Celsius and degrees Fahrenheit.

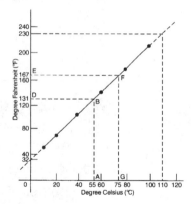

Figure 5.24

(a) To find the Fahrenheit temperature at 55°C a vertical line AB is constructed from the horizontal axis to meet the straight line at B.
The point where the horizontal line BD meets the vertical axis indicates the equivalent Fahrenheit temperature. Hence 55°C is equivalent to **131°F** This process is called *interpolation*.

(b) To find the Celsius temperature at 167°F, a horizontal line EF is constructed as shown in Figure 5.20. Hence 167°F is equivalent to **75°C**

(c) If the graph is assumed to be linear even outside of the given data, then the graph may be extended at both ends (shown by broken lines in Figure 5.24). From Figure 5.24, 0°C corresponds to **32°F**

(d) By extending the graph it is possible to find values outside the given range. This process is called *extrapolation*. From Figure 5.24, 230°F is seen to correspond to **110°C**.

The resistance $R\,\Omega$ of a copper winding is measured at various temperatures t°C and the results are as follows:

$R\,\Omega$	112	120	126	131	136
t°C	20	36	48	58	64

Plot a graph of R against t and find from it (a) the temperature when the resistance is 122 Ω and (b) the resistance when the temperature is 52°C (c) the resistance when t = 70°C.

Activity 5.5

Experimental tests to determine the breaking stress σ of rolled copper at various temperatures t gave the following results.

Stress σ N/m²	8.46	8.02	7.75	7.35	7.06	6.63
Temperature t °C	70	200	280	410	500	640

Show that the values obey the law $\sigma = a\,t + b$, where a and b are constants and determine approximate values for a and b. Use the law to determine the stress when the temperature is 250°C and the temperature when the stress is 7.54 N/cm².

Activity 5.6

In an experiment demonstrating Hooke's law, the strain in an aluminium wire was measured for various stresses. The results were as follows:

Stress N/mm²	4.9	8.7	15.0	18.4	24.2	27.3
Strain	0.00007	0.00013	0.00021	0.00027	0.00034	0.00039

Plot a graph of stress against strain. Find:

(a) Young's modulus of Elasticity for aluminium (given by the gradient of the graph)

(b) the value of the strain at a stress of 20 N/mm², and

(c) the value of the stress when the strain is 0.00020.

Sine and cosine graphs

(i) A graph of $y = \sin A$ is shown by the broken line in Figure 5.25 and is obtained by drawing up a table of values. A similar table may be produced for $y = \sin 2A$.

A	0°	30°	45°	60°	90°	120°	Etc.
$2A$	0°	60°	90°	120°	180°	240°	Etc.
$\sin 2A$	0	0.866	1.0	0.866	0	−0.866	Etc.

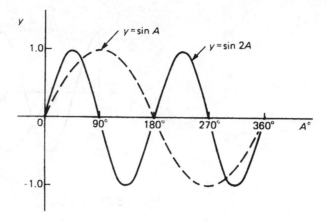

Figure 5.25

(ii)   A graph of $y = \sin \frac{1}{2} A$ is shown in Figure 5.26 using the following table of values:

A	0°	30°	60°	90°	120°	150°	Etc.
$A/2$	0°	15°	30°	45°	60°	75°	Etc.
$\sin A/2$	0	0.259	0.500	0.707	0.866	0.966	Etc.

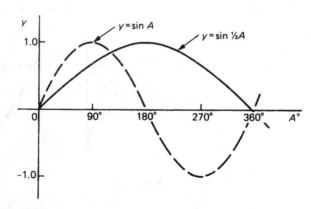

Figure 5.26

(iii)   A graph of $y = \cos A$ is shown by the broken line in Figure 5.27 and is obtained by drawing up a table of values. A similar table may be produced for $y = \cos 2A$. A graph of $y = \cos 2A$ is shown in Figure 5.27.

A	0°	30°	45°	60°	90°	120°	Etc.
$2A$	0°	60°	90°	120°	180°	240°	Etc.
$\cos 2A$	1.0	0.500	0	−0.500	−1.0	−0.500	Etc.

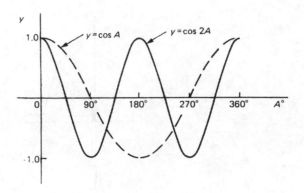

Figure 5.27

(iv)   A graph of $y = \cos \frac{1}{2} A$ is shown in Figure 5.28 using the following table of values:

A	$0°$	$30°$	$60°$	$90°$	$120°$	$150°$	Etc.
$A/2$	$0°$	$15°$	$30°$	$45°$	$60°$	$75°$	Etc.
$\cos A/2$	1.0	0.966	0.866	0.707	0.500	0.259	Etc.

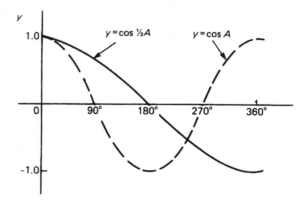

Figure 5.28

Periodic time and period

(i)   Each of the graphs shown in Figures 5.25 to 5.28 will repeat themselves as angle A increases and are thus called *periodic functions*.

(ii) $y = \sin A$ and $y = \cos A$ repeat themselves every $360°$ (or 2π radians); thus $360°$ is called the *period* of these waveforms. $y = \sin 2A$ and $y = \cos 2A$ repeat themselves every $180°$ (or π radians); thus $180°$ is the period of these waveforms.

(iii)   In general, if $y = \sin pA$ or $y = \cos pA$ (where p is a constant) then the period of the waveform is $360°/p$ (or $2\pi/p$ radians). Hence if $y = \sin 3A$ then the period is $360/3$, i.e. $120°$, and if $y = \cos 4A$ then the period is $360/4$, i.e. $90°$.

Amplitude

Amplitude is the name given to the maximum or peak value of a sine wave. Each of the graphs shown in Figures 5.25 to 5.28 has an amplitude of 1 (i.e. they oscillate between +1 and −1). However, if y = 4 sin A, each of the values in the table is multiplied by 4 and the maximum value, and thus amplitude, is 4. Similarly, if y = 5 cos 2A, the amplitude is 5 and the period is 360°/2, i.e. 180°.

Lagging and leading angles

(i) A sine or cosine curve may not always start at 0°. To show this a periodic function is represented by $y = \sin(A \pm \alpha)$ or $y = \cos(A \pm \alpha)$ where α is a *phase displacement* when compared with $y = \sin A$ or $y = \cos A$.

(ii) By drawing up a table of values, a graph of $y = \sin(A - 60°)$ may be plotted as shown in Figure 5.29. If $y = \sin A$ is assumed to start at 0° then $y = \sin(A - 60°)$ starts 60° later (i.e. has a zero value 60° later). Thus $y = \sin(A - 60°)$ is said to *lag* $y = \sin A$ by 60°.

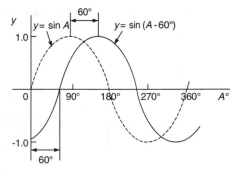

Figure 5.29

(iii) By drawing up a table of values, a graph of $y = \cos(A + 45°)$ may be plotted as shown in Figure 5.30. If $y = \cos A$ is assumed to start at 0° then $y = \cos(A + 45°)$ starts 45° earlier (i.e. has a zero value 45° earlier). Thus $y = \cos(A + 45°)$ is said to *lead* $y = \cos A$ by 45°.

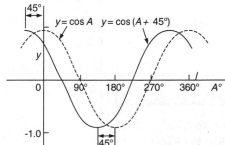

Figure 5.30

(iv) Generally, a graph of $y = \sin(A - \alpha)$ lags $y = \sin A$ by angle α, and a graph of $y = \sin(A + \alpha)$ leads $y = \sin A$ by angle α.

(v) A cosine curve is the same shape as a sine curve but starts 90° earlier, i.e. leads by 90°. Hence cos $A = \sin(A + 90°)$.

Example 5.76

Sketch $y = \sin 3A$ between $A = 0°$ and $A = 360°$

Amplitude = 1; period = 360°/3 = 120°

A sketch of $y = \sin 3A$ is shown in Figure 5.31.

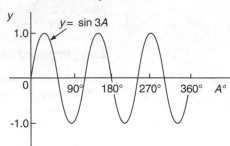

Figure 5.31

Example 5.77

Sketch $y = 3 \sin 2A$ from $A = 0$ to $A = 360°$.

Amplitude = 3; period = 360°/2 = 180°

A sketch of $y = 3 \sin 2A$ is shown in Figure 5.32.

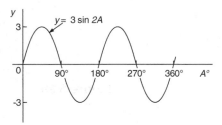

Figure 5.32

Example 5.78

Sketch $y = 5 \sin(A + 30°)$ from $A = 0°$ to $A = 360°$

Amplitude = 5; period = 360°/1 = 360°

$5 \sin(A + 30°)$ leads $5 \sin A$ by 30° (i.e. it starts 30° *earlier*).

A sketch of $y = 5 \sin(A + 30°)$ is shown in Figure 5.33.

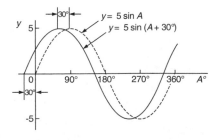

Figure 5.33

Test your knowledge 5.19

1   Sketch $y = 4 \cos 2x$ over the range $x = 0°$ to $x = 360°$

2   Sketch $y = 2 \sin 0.25\,A$ over the range $A = 0°$ to $A = 360°$

3   Sketch $y = 7 \sin(\theta + 15°)$ from $\theta = 0°$ to $\theta = 180°$

4   Sketch one complete cycle of $y = -5 \cos(2t - 60°)$.

Activity 5.7

An electronic signal source produces an output voltage that is described by the following equation:

$v_1 = 15 \cos (\theta + 45°)$

A second signal source produces an output voltage given by the equation:

$v_2 = 10 \sin (2\theta - 30°)$

Sketch a graph showing these two waveforms on a common set of axes. Label your drawing clearly.

For each waveform state:

(a) the amplitude

(b) the period

(c) the phase angle.

Use your graph to determine the values of θ for which the two voltages, v_1 and v_2, are the same. Hence solve the equation:

$15 \cos (\theta + 45°) = 10 \sin (2\theta - 30°)$.

Review questions

1 Evaluate: (a) $2^3 + 3^2$, (b) $3^2 \times 3^4 \div 3^5$, (c) $6^2 \times 2^{-2}$, (d) $16^{1/2}$, (e) $4^{3/4}$, (f) $3^{5/3}$, (g) $9^{-1/2}$.

2 Simplify: (a) $(3^2)^3$ and (b) $(2^3)^{-2}$ expressing the answers in index form.

3 Simplify $\dfrac{(10^3)^2}{10^4 \times 10^{-6}}$ and express your answer in index form.

4 Express the following in standard form: (a) 2173, (b) 0.0124, (c) 0.0000471, (d) 392500.

5 Evaluate $\dfrac{(1.8 \times 10^6) \times (2.5 \times 10^{-9})}{(0.9 \times 10^{-6})} \times (3 \times 10^{-3})$

and give your answer in standard form.

6 Simplify: (a) $a(b + ab^2) - a^2b^2$, and (b) $\dfrac{p^3 q^2 r^4}{pq^{-1} r^3}$

7 Remove the brackets from: $(x - 2)(5x - y^2)$.

8 Factorize (a) $pq + 2pq^2$, (b) $3a^2 - 15a^2b$, (c) $2a^2b - 4ab^2 + 6ab$.

9 Solve $3x - 1 = 4x + 9$.

10 Solve $x^2 + 2x = 8$.

11   Determine the roots of $9x^2 - 49 = 0$ by factorization.

12   Solve $x^2 + 3x - 8 = 0$ using the quadratic formula.

13   A cone has a slant height of 1.2 m and a total surface area of 12 m². Determine the base diameter of the cone.

14   Determine the sine, cosine and tangent of the angle θ in Figure 5.34.

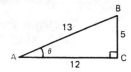

Figure 5.34

15   Determine the value of the angle θ (in degrees) in Figure 5.34.

16   Evaluate, correct to four significant figures: (a) sin 68°, (b) cos 171°, (c) tan 350°.

17   State (a) the sine rule, (b) the cosine rule.

18   Determine the unknown side, y, and angles X and Z of the triangle shown in Figure 5.35.

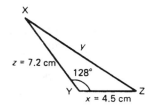

Figure 5.35

19   Determine the area of the triangle shown in Figure 5.35.

20   The following data was obtained in an experiment:

Temperature, T (°C)	60	65	70	75	80	85
Volume, V (m³)		25.0	25.8	26.6	27.4	28.2	29.0

Plot a graph of V against T and use it to find (a) the temperature at which the volume is 28.6 m³, and (b) the volume when the temperature is 55°C.

21   Sketch graphs showing (a) $y = 3 \cos 3x$ over the range $x = 0°$ to $x = 180°$, and (b) $v = -2 \sin(\theta/2 + 45°)$ over the range $\theta = 0°$ to $\theta = 360°$.

Unit 6 Design development

Summary

This unit will help you to acquire the skills required to develop and communicate designs, and successfully produce design assignments either alone or as a team. It will also complement your communication skills.

The unit is divided into three elements. The first deals with design briefs for an electro-mechanical engineered product and an engineering service. The second is concerned with producing and evaluating design solutions for an electro-mechanical engineered product and an engineering service. The final element deals with the technical drawings used to communicate designs for engineered products and engineering services.

Before we start it is necessary to establish what types of an engineered product and engineering service you should be considering. The engineered product should be an *electromechanical device*. For example, a remote controlled garage door or an electronically activated locking system, or any design that is a combination of electrical or electronic components used with a mechanical device.

The service chosen should involve *installation or maintenance*. For the purposes of this unit, maintenance is assumed to include repair. For example, a computerized network to control factory production or a distribution service for one of the utilities such as gas, electricity etc. Try to think of some service that you are familiar with.

Think hard about the types of engineered product and engineering service you choose. In both cases, try considering something not too complicated but at the same time one which satisfies the requirements of your assignments. We will first look at a description of design, then consider the requirements of a design brief. The principles of a design brief will equally apply to both an engineered product and to an engineering service.

Design

We have always designed things, it is a basic characteristic of us all to design to meet our needs. It does not require a special ability to design, as often a craftsman may make an item without the need of drawings or modelling before the completion of his product. In industry, where many items are to be manufactured, the activities of designing and manufacturing are quite separate.

For the purposes of this unit, engineering design and manufacture are considered to be where the process of manufacturing a product cannot proceed until the design of it is complete, this is the usual procedure for the manufacture of the majority of factory made engineering products.

In some cases, e.g. in the automobile industry, the time taken to design a car can be several years, whereas the time to manufacture each individual vehicle may be measured in hours.

If a product cannot be manufactured until the design is finished, then at least it is clear what the design process needs to achieve – it must provide a complete description of the product that is to be manufactured, with almost nothing left to the discretion of the production team. The shape, materials, dimensional tolerances and finishes, etc. will all be specified by the designers.

The essential design activity then, is the producing of a set of drawings which completely satisfy the customers or marketing design requirements, and equally as important, are suitable for communicating the design manufacturing requirements to the production department.

Therefore, the drawings will play a very important part in the design and manufacture operations and are controlled by a specification to enable a common standard of presentation to be achieved (British Standard 308, Engineering Drawing Practice, is the usual accepted standard). Later, we will discuss the requirements of the drawing 'design package' required to manufacture a product.

Basically, design is about 'problem finding'. If this seems odd, the design process involves addressing the problems identified by the customer and the marketing department. In many situations, problem finding is more important than problem solving!

Einstein said, *'The formulation of a problem is often more essential than its solution, which may be merely a matter of mathematics or of experimental skill. To raise new questions, new possibilities, to regard old problems from new angles requires creative imagination and a real advance in science.'*

The design brief

Here we will consider what is a design brief, this can only be decided after some thought has been given as to the design changes or requirements. In other words what do you expect the design to achieve? If you have completed the exercise on Einstein's statement, you may have produced a design brief of sorts.

Design problems usually come by a statement to the designer from either a customer requirement or from the company management. These statements are referred to as design 'briefs', they can vary considerably in both content and form.

The normal 'brief' statement from the company management is probably to improve the design of an existing product that is familiar to the company designers, it may be related to performance, size or weight, etc. or maybe as a redesign to reduce manufacturing costs to keep the product competitive on the market.

Other design problems occur by statements such as 'it would be nice if ....', or 'is it possible to have ....'. Often this type of comment would be expected from a product user to the marketing department, who in turn would approach the designers to find whether the customer suggestions could be made feasible.

Perhaps the greatest task given to designers in recent times was the brief' given by the USA President Kennedy in 1961 whose brief was, *'Before the end of the decade, to land a man on the moon and bring him back safely.'*

The only constraint in this brief being one of time, so in this case the designers had a fixed goal, only one constraint and large resources of money, materials and people to work on the project. Think of all the problems and sub-problems that must have occurred.

It is important to note that a design 'brief' does not specify what the design solution will be, and there is no way of proceeding from the stated 'brief' to a proposed solution without designing.

What is generally expected from a design brief would be certain constraints stated by the customer and legal constraints such as the product safety, also standards and legislation relevant to the product. This is applicable to an engineering service equally as well.

The considerations that should be taken into account to produce a design brief are: (a) customer requirements, (b) standards and legislation, and (c) constraints. We shall consider each of these factors in turn.

Customer requirements

Customer requirements include:

- functional considerations (technical specification)
- ergonomic considerations (including ease of use and adaptability to suit different users)
- aesthetic considerations (details on styling, general appeal, range of colour options, etc.)
- quality (type of materials, reliability and expected life)
- cost (initial purchase price)
- whole life cost (ease and cost of repair, the need for routine maintenance and the cost of spare parts and consumable items, as well as purchase price)
- quantity (prototypes and production costs can be directly related to the quantities involved)
- size and weight (this is often is of prime importance to the customer)
- time-scale (involves creating a realistic planning and development schedule to meet customer requirements)
- tolerance.

Standards and legislation

The considerations that may apply under this heading include:

- Health and Safety legislation
- codes of practice
- conventions

- British Standards
- European Union directives
- International Standards.

Standards and legislation varies widely according to the type of product or service. You are expected to be aware of any legislation relevant to the product or service you are studying, you are advised to consult the latest BSI *(British Standard Institute)* catalogue for the relevant standards and information you need.

Constraints

The constraints that may impinge on the design brief include:

- technological
- resources (labour, materials, plant)
- environment
- cost.

In most cases it is necessary to work well within the bounds of existing technology while recognizing that some design briefs may demand technological development before implementation.

The brief need not necessarily apply to a new invention, it is quite acceptable to redesign part of a product, for example, to reduce the number of components or take advantage of new materials and technology. Having checked the customer requirements, standards and legislation and the constraints imposed on you, you should now be able to produce a design brief.

A design brief is a description of the customers needs, so it is vital to get the design brief correct. You should discuss the brief with your 'customer', ideally someone in industry, more likely your lecturer or tutor, using the listed considerations as a 'ticking list'. Make absolutely certain that the brief *clearly and unambiguously* states the customer requirements!

Feasibility studies

The next stage is a feasibility study of the design 'brief'. This entails an in-depth study of the brief to establish whether a feasible design is possible to satisfy all the requirements the customer expects, it is at this stage all sorts of problems need consideration. Sometimes these problems will require a compromised solution agreed between you, the design team and the customer.

By the time you and your team have finished the feasibility study and found proposed solutions to satisfy the design brief, many ideas and solutions are experienced, remember though at this stage the task has only been to see whether the design brief is feasible, and probably very little real design work has taken place, apart from investigations and design possibilities.

The problems most generally encountered in engineered products are perhaps those associated with weight or size, and other problems

---

Test your knowledge 6.2

Acme Tools wish to extend their range of small production tools to include an adjustable jig for holding a Eurocard printed circuit board during inspection and assembly. In use, the jig is to be clamped to a workbench and it should rotate and swivel to allow inspection of the printed circuit board from any angle.

What items would form an initial design brief for this product? What questions would you need to ask Acme Tools?

may occur by not being able to keep pace with changes in technology, especially electronic devices, and you could find your design is outdated 'before it has left the ground'.

When you are confident that you will be able to further the ideas expressed in the design brief from the customer and you are satisfied your feasibility study shows you would be capable of handling the design, the next stage is to submit to your customer a design proposal.

Design proposals

By proposing solutions, a greater understanding of the overall design problem will be gained which in itself will show many sub-problems. The design proposal can take a considerable amount of time and money to investigate and produce, it will need to convince the customer that you are, as a company, capable of solving the problems of design, and have the expertise and qualified staff to do so.

On an elaborate equipment that requires knowledge that is close to the edge of technology, for instance, a ship's complete radar system, the design proposal can cost thousands of pounds and take months to complete. If the contract demands a competitive tender, as it almost certainly would, then even more time will be spent to estimate the cost of the contract, which is a difficult task as the final design specification has yet to be agreed.

Fortunately your team project need not be as elaborate, but you will still need to analyse your customer's design brief and your own feasibility study before any design proposal can be produced.

The design proposal sent to the customer basically consists of:

* a provisional design specification
* proposed overall contract costs
* delivery dates
* servicing and maintenance
* any legal requirements.

The above is very much simplified but you will need to consider each of them to enable you to write the design specification for your own project.

Design specification

Up to this point only a few staff have been fully employed in analysing the customer brief and submitting the design proposal. Very little real detail design has taken place and only a broad view of the customer requirements has been formed. Assuming that the design proposals have been agreed by the customer, your company will then be given the task of producing a detailed design specification.

A basic project design team would be formed to look at the problems in greater detail and write the design specification which will be similar to the proposal but with more detail of the design targets and time-scales that must be achieved.

In the case of your own project, you will probably be working as part of a team of three or four students. Collectively you will act as the project design team. Furthermore, you will have multiple tasks to handle in much the same way as a real team in industry.

Some of the departments and personnel that will form the nucleus of a design team, will have already been involved with the project right from the start (i.e. at the design brief and design proposal stages). These people will therefore have prior knowledge of some of the problems they will encounter and they will thus be ideally suited to producing the design specification, which will hopefully meet with approval of the customer.

We will now briefly look at the team required to handle a design specification in engineering and this may give you an insight into possible career paths.

In small design and manufacturing companies, one person may take responsibility for more than one task (as you will, in working within a team on your project). As an example, the Company Managing Director may also take on the role of Financial Director and perhaps also be directly responsible for staff recruitment. But for our purpose we will assume the company is of reasonable size and each task is allocated to an individual (Figure 6.1).

At the Company Board level, both the financial and research and design directors will need to agree the necessary costs of the

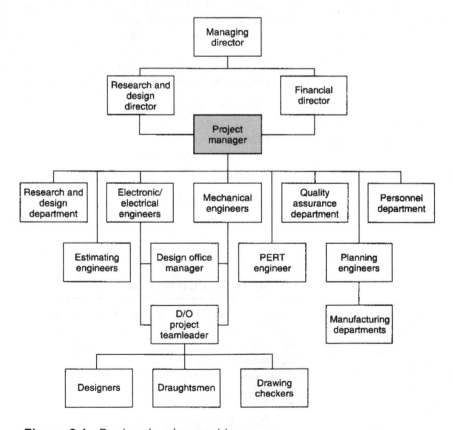

Figure 6.1 *Design development team*

intended specification. The design team, usually headed by a project manager, will basically consist of electronic/electrical engineers, mechanical engineers, designers and draughtsmen in the design office. In addition, planning engineers will consult with the team and plan a task and time-scales chart, the chart itself will in all probability be generated by a PERT engineer (Programme Evaluation and Review Technique).

Planning and manufacturing departments will need to assess available facilities and whether any special machinery or test equipment should be acquired and the quality assurance department should check that the company is capable of dealing with the required standards of the design specification or brief.

The personnel department may be tasked to find staff with capable skills, experience and specialized knowledge if the project requires it. When the tasks of all departments have been completed the design specification can then be written. When this specification is agreed with the customer, a contract will then be drawn up, stating the cost to the customer, date of delivery, servicing contract details and any legalities that may apply to the main contract. There may also be a penalty clause (for non-delivery of goods on the agreed date) written into the contract.

As the project develops and unseen problems occur the contract may require 'fine tuning' by both the contractor and the customer. In the electronics industry, the continuous improvements to available devices sometimes make design out-dated as the project develops so certain areas may be subject to renegotiation with the customer.

Whilst considering the design proposal, many people were involved and they will almost certainly form the basis of the final project design team. Now each will be responsible for their own department's role in the progress of the project. The main responsibilities of each will be to monitor the allocated costs to their department, to establish time-scales of the different tasks and to formulate the design objectives.

Matrix management

In setting up the project team a 'matrix organization' is often adopted in which the project manager is allocated teams from different departments in the organization. The team members report in the normal way to their line managers but have a reporting link to the project manager as well. For instance the drawing design office team are responsible to the drawing office manager or chief draughtsman but report the progress of the drawing package to the project manager.

The advantages of adopting a matrix structure are:

1 Flexible deployment of the company resources.
2 Effective availability of specialized knowledge.
3 The ability to run more than one project at a time.
4 Better career paths in the specialist areas.

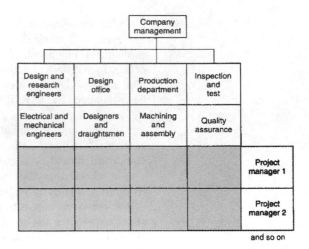

Figure 6.2 *Examples of matrix management*

In Figure 6.2 we can see that the various project managers can have access to the expertise of all the other departments needed to execute the research, design and production required, without having the need to set up a unique team for each function of design and manufacture of his own individual project.

It often happens that the project manager has staff working in his team that in the company job structure are at a higher level than himself, but these staff are usually highly qualified experts in their field and they are available for the needs of any company project.

The design specification can take a considerable amount of time and money to investigate and produce, in industry it would need to convince the customer that you are, as a company, capable of solving the problems of design, and have the expertise and qualified staff to do so.

Before continuing let's assume you have been given your Design Assignments and that you need to complete them in say, 24 weeks, and you have, as a class, been divided into teams of three or four students to produce the assignment. Each of you will need to be tasked to investigate and solve the problems involved with your design, and *all of you will need to meet the time-scales*. You will need to produce an action plan.

What is an action plan?

Basically, an action plan is a chart of tasks that are time related and show the team responsibilities to the design, it will determine how long relative tasks should take. It is a vitally important design document and must be produced at the very beginning of your assignment. You should, of course, design your own action plan to suit your requirements, but Figure 6.3 will give you some idea of how it should look. You may of course need to adjust the time-scales and functions to suit your assignments, but in principle the chart will be similar. Planning is an essential part of the success or otherwise

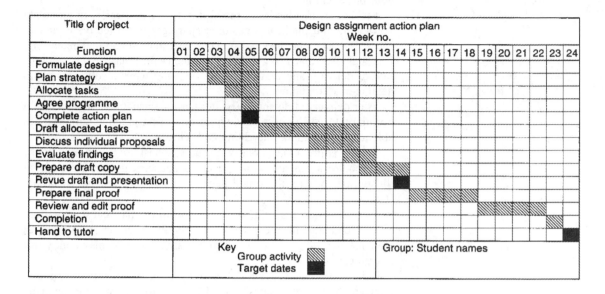

Figure 6.3 *Example of an action plan*

of an assignment, without careful planning, you as a team will find difficulty in achieving the target dates. It will be up to you to interpret the assignment requirements, but of course your tutor is available to assist you, but try to plan your actions first.

It is very important to keep the dates you commit yourself to as 'set-out' in the plan, also any dates set by your tutor must be adhered to, such as completion of action plan, revue of draft and verbal presentation and, of course, the final date to hand the project to your tutor. Failure to achieve these dates could lose you marks.

The action plan may look simple enough, but there is need to investigate many areas of design to succeed in your assignment, so seriously consider what you want your design to accomplish and adhere to the plan.

Before you start on your own design ideas, practice writing a design specification, consider what is required and work as a team, or preferably ask the tutor to involve the whole class as this will help you understand the needs for your own assignment.

Presenting your work

Whether you are individually or as a group presenting your projects and assignments, each of you need to try to observe the following methods of delivery. Your style, mannerisms, speech, gestures, eye-to-eye contact with your listeners, clarity of expression, appearance and 'personality' can make a considerable impact on them.

You will, whilst presenting your assignments, be the sole focus of attention for most of the time; your style of delivery can result in acceptance and assimilation, or rejection, of your assignment presentation content. Your 'expressiveness', i.e. your obvious enthusiasm for your project, your perceived desire to communicate and your ability to generate listener interest, can 'make or break' your presentation.

Techniques involved are: the ability to speak clearly, to modulate voice tone and pitch, use gestures sparingly but effectively, and to speak at a pace that does not prevent assimilation and understanding. It is suggested that 110 words per minute (wpm) may be the 'normal' rate of delivery and that a rate beyond 200 wpm results in a rapid decline in assimilation.

Your delivery style should be neither too casual nor too much of a boring oratory. The best way would be to first practice until you are able to sound natural, interesting and relaxed.

Timing is important along with the emphasis of key points, these may require an intensity of speech, a pause, a gesture or a visual illustration. A carefully timed pause can serve as a signal for a key statement.

Some mannerisms (e.g. of voice, posture) may cause irritation, even offend and alienate, creating a difficulty in communicating with your listeners, if you are aware of a mannerism that may annoy, try to eradicate it.

It is vital that you should convey a genuine enthusiasm and interest. Non-verbal cues, facial expressions, eye contact (the absence of which can be interpreted by the listeners as nervousness, fear or lack of interest), and use of hands is all important.

Where your presentation involves the use of visual aids, they should be prepared in advance and in the order required, so continuity of presentation runs smoothly.

Timing

It is very important to adhere to the time allowed to present your assignment, therefore you individually or as a group need to plan who will be responsible for which part or parts, you may only be given as little as 10 to 15 minutes to complete your presentation, so it is necessary to draw up a simple 'plan of action' to ensure the smooth presentation of your assignment. I would suggest you take into account the following in the sequence shown:

1 Introduction of yourself or group, it may seem unnecessary to do this to a class of colleagues but you may well be a stranger to the lecturer in charge.

2 State clearly the objective of your project or assignment and, in the case of a 'Design Assignment', briefly why or how you finally chose the subject for design.

3 Now talk about how your ideas came to fruition including marketing, costs, manufacturing and customer guarantees, if applicable, these may be split as tasks to be dealt with by other team members, in which case they need to present them individually.

4 Be prepared to answer questions, preferably at the end of your your presentation, think and answer carefully. If you are unable to do so, say that you will investigate further and let the questioner know as soon as possible.

5 Finally, thank your listeners, especially those that may have contributed during question time.

6 Above all, *be confident in yourself.*

Activity 6.1

You have been commissioned by an environmental group to produce a design proposal for a prototype of an environmentally friendly car. Use your library to carry out some research on this topic and produce an initial design proposal for a vehicle that will:

(a) be suitable as a town car
(b) carry one adult and two young children, plus family shopping
(c) must be easy to park
(d) must be inexpensive to run.

Present your findings to the class using appropriate handouts and visual aids. Allow plenty of time for questions at the end of your presentation.

Design solutions

Having dealt at some length with design briefs and design specifications, we can now consider how you should go about handling a design specification (assuming you are now prepared to go ahead with your proposed design). At this stage, it is worth remembering that the design specification is not a design solution.

The design path

Having established your design team that now has a reasonable insight into the design requirements of the project specification, we can set up a design programme of the project from start to finish. However, before designing begins we need to consider what design strategy should be adopted.

Design strategies

What is a design strategy? It is basically having a design method and it consists of two things:

1 A *framework of intended actions* within which to work
2 Some form of *management control function* to enable you to adapt your actions as the problem unfolds.

Using a design strategy may seem to divert effort and time from the main task of designing, but this may not be a bad thing, as the purpose of a design strategy will make you think of the way design problems will be dealt with. It also provides you with an awareness as to where the design team is going and how it intends to get there.

The purpose then of having a strategy is to ensure that design activities are realistic with respect to the constraints of both time and resources within which the design team has to work. In a manufacturing company the most used strategy is a sequence of actions that have previously been applied to an already existing product.

For instance, to design a variation of an already designed electromechanical device or engineering service, the strategy most likely to be used for the new variation will be the same tactics and design methods used for the previous design. This would therefore be making use of a 'pre-established' strategy. The relevant tactics would be drawn from conventional techniques and rational methods already familiar to the design team. This type of strategy applies to innovative designs.

It is not always possible to have a Design Strategy, as would be the case in research design situations but having no particular plan of action would be a type of strategy in itself. This could be referred to as an 'inventive' strategy. The type of final design may be purely inventive, where no previous market exists.

Often the designers may not know when or what the final outcome may be, although hopefully they may achieve some degree of success in designing a material, product or engineering service that can be commercially exploited. The relevant tactics would be mainly creative.

The two strategies mentioned are extreme forms. In all probability, most designs require a compromise between the two, certain parts of the project design may need the inventive strategy if it calls for unknown areas of engineering design.

The 'pre-established' strategy is predominantly a convergent design approach, whereas the 'inventive' strategy is predominantly a 'divergent' approach. Usually the aim of a design strategy is to converge onto a final, evaluated and detailed design.

Sometimes in reaching that final design it may be necessary in some areas of the design to diverge, so as to widen the search for new ideas and solutions. Therefore the overall design process is mainly convergent, but has elements of divergent thinking.

Convergent thinkers are usually good at detail design, and evaluating and choosing the most suitable solution from a range of options. On the other hand, divergent thinkers are best at conceptual design problems and are able to produce a wide range of alternative solutions.

The design team needs both types of thinkers for a successful design, but generally engineering promotes and develops only convergent thinkers.

There are basically six stages in a design action framework that should be adopted right from the start of the design process:

1 Clarifying objectives.
2 Establishing functions.
3 Setting requirements.
4 Generating alternatives.
5 Evaluating alternatives.
6 Improving details.

Now let us look in more detail at the stages that make up the design process and expand the meaning of each:

1 Clarifying objectives

This is stage would in all probability have been dealt with during the design brief and design proposals stages, when the objectives of the design would be stated. But you would still need to clarify them before design can commence.

Aim
To clarify the design objectives and sub-objectives, and the relationships between them.

Method
This is best achieved by discussions with the design team and questions to the customer.

2 Establishing functions
Although the customer may have specified the functions expected from the design the designer may find a more radical or innovative solution by reconsidering the level of the problem definition. He or she may be able to offer the customer a better solution to the functional problems of the design in excess to the expected at no extra cost.

Aim
To establish the functions required.

Method
Break down the overall function into sub-functions. The sub-functions will comprise of all the functions expected within the product.

3 Setting requirements
Design problems are usually all set within certain limits, these limits may be cost, weight, size, safety or performance, etc. or any combination of them.

Aim
To produce an accurate specification of the performance of the designed product.

Method
Identify the required performance attributes, these may well have been considered at the design feasibility study stage.

4 Generate alternatives
Even if you think you have a good design solution always look further, if time and costs permit for an alternative solution, hopefully better than the one you thought would be an ideal design.

Aim
To have a choice of solutions to allow comparisons of ideas in solving the design requirements.

Method
It would help to draw several design layout drawings to enable discussions to take place with the design team and the customer.

5 Evaluate alternatives

When some alternative design proposals have been thought about and maybe some design layouts have been produced, the evaluation of the alternatives can be discussed.

Aim

To evaluate the alternatives, choose the ones that satisfy the customer requirements and are good value to him.

Method

Compare the value of the alternative design proposals against the original proposal agreed with the customer on the basis of performance.

6 Improving details

There are mainly two reasons for improving details, they are either aimed at increasing the product value to the customer or reducing the cost to the producer.

Aim

To increase or maintain the value of the product to the customer at the same time reducing its cost to the producer.

Method

This can be approached using two methods, one is called value engineering and the other value analysis.

Value engineering and value analysis

Value engineering.

A great deal of design work in practice is concerned not with the creation of radical new design concepts but with making modifications to existing product designs. Value engineering is used for this purpose, but it is difficult to establish without a prototype or model of the design project, as it basically requires listing of all the separate components of the product so that the function of each may be identified and evaluated. For instance, it is often found that further standardization is possible from one unit assembly to another that was not previously obvious from the design drawings. This often results from the complexity of the design break-down and the team not being aware of the different approaches to the solving of problems within the same team. Hence the team leader should keep an eye on the design as it progresses and check that the team tries to standardize parts and components wherever possible, by doing this he will make a valuable contribution to the success of the product.

When carrying out a value engineering exercise, it will often be found that some manufactured parts are over-designed for their required function or, by using different manufacturing techniques, improvements can be made to the reliability of the product. It is important to remember though, any design change due to the value engineering exercise should not diminish the customer requirements of the product and, if anything, should enhance it.

Some manufacturers may purchase a competitor's product to enable them to subject it to the value engineering method, and afterwards design and market an improved competitive version. This is one way of learning without resorting to industrial espionage.

It does not take much imagination to see how this method has been applied to industries concerned with the motor, electronics and domestic appliance markets.

So we are looking for ways of reducing costs without reducing value or adding value without adding cost. This design stage requires both critical and creative thinking, it means critically looking at the design as it is and creatively thinking how it could be.

Care must taken not to change the design concept purely for the sake of change as can sometimes happen. The aim must always be to achieve high value functions with low cost components.

A checklist for cost reduction guidelines can be as follows:

1 Standardize Can parts be standard rather than special?
2 Modify Is there a satisfactory cheaper material?
3 Reduce Can the number of components be reduced?
Can several components be combined into one?
4 Simplify Is there a simpler alternative? Is there an easier assembly sequence? Is there a simpler shape?
5 Eliminate Can any function and therefore its components, be eliminated altogether? Are any components redundant?

Value analysis

Value analysis can perhaps be best described by a product that meets the needs of the customer at a competitive price. This can be achieved in a similar manner to value engineering but with greater consideration to the customer/market. This does not mean paring the price down to rock bottom, as the product in some cases may be designed to suit popular demand, take for instance a car radio tuner, push button tuning is more popular than knob tuning, although the cost may be considerably higher, the customer may accept this as better value.

No modifications to a product should diminish the customer's requirements. Modifications should seek to:

• improve a product by improving its performance
• reduce its weight
• lower its cost
• enhance its appearance, and so on.

But at all times keep your customer (and hence your market) in mind. In other words Value Analysis really means, '*is your designed product meeting the demands of the customer at a reasonable cost?*' Can you enhance your design within a competitive price range?

It can be seen that Value Engineering and Value Analysis must go hand in hand as both have an equally important role in manufacturing costs and the product that is delivered to the customer.

We have now looked at a system that could be adopted as a complete 'pre-established' strategy. It consists of a framework covering the design process from customer requirements through to detail design. The other important function of a successful design strategy is that it must be well controlled by good management.

A designer working alone will still require to manage his own design strategy, but if it involves a team of designers, either the team leader or the whole team collectively must review the progress of the design and if necessary amend the strategy, time can be wasted if the design strategy is not organized.

These are suggested rules for keeping the design strategy under control:

- Be sure to keep all objectives in mind; in designing it is impossible to have only one set of completely fixed objectives.
- A creative resolution of a design problem often means changing some of the earlier objectives.
- The design strategy should be kept under continuous review, the aim is to solve the design problem in a creative and competent way and not follow a path leading to nowhere. If no progress is being made, review the strategy.
- Involve others in the team, they may see the problem in different ways, and may be able to suggest a different path and change the way of thinking of a solution to the problem.
- Keep all files and sketches throughout the project design stages, jot down any ideas that come to mind to be possibly used at a later stage, even when working on different aspects of the project.
- *Never* throw away any sketch or layout drawings until well after the design is finished and proven.

Activity 6.2

From the list given below, choose a product you are familiar with and then:

1 Think of ways of reducing its manufacturing costs without reducing its value (value engineering).
2 Improve its function to give you more value within a reasonably competitive price (value analysis).

Present your results in the form of a written memorandum to the product's Design Team Leader.

(a) an electric drill
(b) a garden thermometer
(c) a set-top TV aerial
(d) a battery tester
(e) a cycle lamp
(f) a tyre pump
(g) a battery charger.

Consider a number of different ways of achieving the tilt and swivel action required in the Acme Tools Eurocard jig (see test your knowledge 6.2). Sketch possible design solutions and compare these with those produced by other students. Evaluate each of the solutions produced and see if you can identify the best solution. List the criteria that you used in the evaluation process.

Almost certainly you will need to communicate your design by drawing it, as this is by far the easiest way to describe your intention, in fact to verbally describe a design can cause confusion as the persons listening may well imagine something totally different to the idea as you perceive it.

A knowledge of a variety of methods used for graphical communication is essential if you are to successfully submit your design ideas. In particular you should be able to:

- select appropriate graphical methods to be used for communicating engineering information
- produce scale and schematic drawings for engineering applications
- interpret information presented in engineering drawings.

Generally you will need sufficient drawings to fully explain your design concept, this means an ability to draw the following, if required:

1	Layout drawings	The original sketches and drawings required to show your design proposals
2	Detail drawings	Dimensioned drawings of any manufactured parts
3	Assembly drawings	Showing how the project should be assembled
4	Item lists	Listing all drawn and 'bought out' parts required to make the final assembly

Although drawings are preferred to be drawn to British Standard BS 308, you may find sketches drawn in good proportion are acceptable, you will need to ask your tutor. Also note that initial design concepts are often made up from rough sketches.

Remember to state the materials and finishes required on any of the parts to be drawn and be careful with your dimensioning and tolerancing. It is important to ensure that your drawings or sketches can be understood easily and without confusion.

If you have experience in using CAD (Computer Aided Design) you will know how much easier it is to draw the Assemblies and Details, etc. required in a drawing package.

To help you understand the requirements of a drawing package for a product, it is good idea to look at what a production engineer may expect from the drawings prior to them being issued for manufacture.

In a class or group situation, take it in turns to describe common objects (some examples are given below) without drawing them, waving your hands about, or saying what they are used for. See how quickly others in the group can identify the object!

(a) a paper clip
(b) a hairpin
(c) a drawing pin
(d) a pencil
(e) a mains fuse
(f) a spark plug
(g) a cigarette lighter
(h) a gas lighter
(i) an oil filter
(j) a barometer.

Activity 6.3

You have been commissioned to write an article for your local paper. The article is to appear within a feature on 'New Technology' and you have been asked to explain, in simple terms and using diagrams, how a how computer aided manufacturing system can be used within an engineering company. Produce your article in word processed form (using no more than 1,500 words) and use a CAD or other drawing package to produce original artwork to be used for the diagrams.

Before design drawings are issued to the production department for manufacture, it is important to obtain approval for each drawing from a qualified production engineer, this can considerable help the cost of manufacture, as he or she will be able to advise as to the suitability of the product for the various methods available to produce the individual machined parts. He or she should also check all aspects of the drawings effecting methods of production, datum's, tolerances, etc.

Material suitability

Is the specification correct? Is the material machinable and/or weldable if required? Could machining time be saved by using stock sizes? (check raw material tolerances)

Dimensioning

Check that there are sufficient dimensions to manufacture the item and the drawing can be clearly understood. Dimensioning drawings from left to right is good drawing office practice, check that no dimensions are left to machine shop calculation. Ensure inside and/or outside radii are stated.

Datums

Datums are preferred if made from vertical and horizontal edges or a datum hole, the fewer datums the better (holes can be used for tooling purposes).

Tolerances

Note the tolerances specified, if in doubt enquire whether any tight tolerances stated are justified. Wider tolerances could save considerably on manufacturing costs (often there can be a misuse of geometric tolerancing).

Machined finishes

Surface texture is often specified on a drawing, check the necessity of any very fine finishes shown, by establishing the function of the component part.

Heat treatments

Check the heat treatment specification if quoted, is correct for the material.

Coated finishes

Check the specified coated finish is applicable to the material used.

Machine processes

Could the shape be slightly changed to allow for easier machining, if the answer is yes, consultation with the designers will be necessary.

Assemblies and sub-assemblies

Check whether the number of parts could be reduced by using

standard parts, re-design or machining from the solid. Check the build up of tolerances on assemblies.

Other points

Do the final assemblies satisfy any interchangeability requirements? Is the item commensurate with the layout or scheme previously vetted in conjunction with the designers?

Notes:

1 *At the layout stage*, agreement must be achieved with regard to the technique and method of manufacture, materials, critical dimensions, environment, conditions of use, etc. These should *all* have been satisfied and agreed.

2 Contentious points that may arise must be solved by arbitration with the Chief Designer and the Chief Production Engineer. Do not fall into a position where design is accepted and becomes a 'challenge' to produce!

3 Log all information where contentious points arise, keep any 'marked up' prints, this is for your own protection at a later date.

4 'Sign off' the drawings only when you are satisfied that all criteria have been covered to the best of your knowledge, practice and experience.

Co-operation between designers and production engineers from the layout stage to the final assembly is essential for a well made product.

As the Production Engineer's involvement progresses with knowledge of the detail of a new design, he or she will sometimes need to contact special material suppliers or may require the services of outside contractors for machining purposes, heat treatments or surface finishes, he or she will need to alert the Quality Assurance Department of the requirements so that these outside services can be given quality assurance approval if they are not already listed.

Activity 6.4

Acme Tools have asked you to produce a set of sketches and detail drawings to support your design proposals (see Test your knowledge 6.2 and 6.4). For the Eurocard jig, produce:

(a) a sketch
(b) a layout drawing
(c) a detail drawing
(d) an assembly diagram
(e) an item list.

Present your work in the form of a portfolio of drawings.

Review questions

1   List THREE main considerations that should be taken into account when producing a design brief.

2   Explain the purpose of a design proposal.

3   Describe TWO different design strategies.

4   Explain the difference between a convergent design strategy and a divergent design strategy.

5   Briefly describe the role of the project manager in the evolution of a new design.

6   Explain TWO desirable outcomes of a value engineering exercise.

7   Identify the roles of the members of a typical engineering design team.

8   Explain the advantages of a matrix organizational structure in relation to engineering design projects.

9   Produce an outline design specification for a portable workbench suitable for the DIY enthusiast.

10   Explain the relationship between cost and quality levels in relation to an engineering design.

11   Describe FOUR different types of diagram that may be used as part of an engineering design solution. Illustrate your answer with sketches.

12   Briefly explain the use of computer systems in arriving at an engineering design solution.

13   Identify THREE constraints that are likely to have an impact on a design brief. Illustrate your answer by giving a typical example.

14   Explain what is meant by a feasibility study in relation to a design brief.

15   Describe the sequence of stages in arriving at a design for an electric screwdriver suitable for home use. Use a diagram to illustrate your answer.

Useful formulae

Area, length and volume

Shape	Feature	Formula
Rectangle	Area Perimeter	$A = l \times b$
Parallelogram	Area	$A = b \times h$
Triangle	Area	$A = \frac{1}{2} \times b \times h$ $A = \frac{1}{2}\, bc \sin A$ $A = \sqrt{s(s-a)(s-b)(s-c)}$ Where $\quad s = \dfrac{a+b+c}{2}$
Trapezium	Area	$A = \frac{1}{2} \times h \times (a + b)$
Circle	Area Circumference	$A = \pi r^2 = \pi d^2/4$ $L = \pi d = 2\pi r$
Semicircle	Area Circumference	$A = \pi r^2/2 = \pi d^2/8$ $L = \frac{1}{2}\pi d = \pi r$
Sector of a circle	Area	$\theta/360 \times \pi r^2$ (θ in degrees) or $\frac{1}{2}r^2 \theta$ (θ in radians)
Cone	Surface area Volume	$A = 2\pi rh + 2\pi r^2 = 2\pi(h + r)$ $V = 1/3\ \pi r^2 h$
Frustum of a cone	Curved area Total surface area Volume	$\pi l(R + r)$ $\pi l(R + r) + \pi R^2 + \pi r^2$ $1/3\ \pi h(R^2 + Rr + r^2)$
Sphere	Surface area Volume	$A = 4\pi r^2$ $V = 4/3\pi r^3$
Hemisphere	Surface area Volume	$A = 2\pi r^2$ $V = 2/3\pi r^3$
Pyramid	Surface area Volume	Sum of areas of the triangles forming the sides plus the area of the base $V = 1/3Ah$
Prism	Surface area Volume	Sum of longitudinal area plus area of the two ends C.s.a. $\times$ length

Mechanics

$$\text{Density} = \frac{\text{mass}}{\text{volume}}$$

$$\text{Stress} = \frac{\text{force}}{\text{area}}$$

$$\text{Force} = \text{mass} \times \text{acceleration}$$

$$\text{Strain} = \frac{\text{change in length}}{\text{original length}} \times 100\%$$

Motion

$$\text{Speed} = \frac{\text{distance}}{\text{time}}$$

$$\text{Acceleration, } a = \frac{v_2 - v_1}{t}$$

$$\text{Distance, } s = v_1 t + \tfrac{1}{2}at^2$$

$$\text{Final velocity, } v_2 = v_1 + at \quad \text{or from} \quad v_2^2 = v_1^2 + 2as$$

Energy

$$\text{Potential energy} = m\,g\,h$$

$$\text{Kinetic energy} = \tfrac{1}{2}\,m\,v^2$$

Heat

$$\text{Quantity of heat} = \text{mass} \times (\text{specific heat capacity})$$

Trigonometric ratios for a right–angled triangle

$$\text{Sine} = \frac{\text{opposite}}{\text{hypotenuse}}$$

$$\text{Cosine} = \frac{\text{adjacent}}{\text{hypotenuse}}$$

$$\text{Tangent} = \frac{\text{opposite}}{\text{adjacent}}$$

Pythagoras' Theorem

$$(\text{hypotenuse})^2 = (\text{opposite})^2 + (\text{adjacent})^2$$

Sine rule

$$\frac{a}{\sin A} = \frac{b}{\sin B} = \frac{c}{\sin C}$$

Cosine rule

$$a^2 = b^2 + c^2 - 2bc \cos A$$

Ohm's Law

$$V = I \times R \quad \text{or} \quad I = \frac{V}{R} \quad \text{or} \quad R = \frac{V}{I}$$

Electrical power

$$P = I \times V \quad \text{or} \quad I = \frac{P}{V} \quad \text{or} \quad V = \frac{P}{I}$$

$$P = I^2 \times R \quad \text{or} \quad P = \frac{V^2}{R}$$

Resistivity

$$R = \frac{\rho l}{A}$$

Resistors in series

$$R = R_1 + R_2$$

Resistors in parallel

$$\frac{1}{R} = \frac{1}{R_1} + \frac{1}{R_2} \quad \text{or} \quad R = \frac{R_1 \times R_2}{R_1 + R_2}$$

Sine wave voltages

$$v = V_m \sin(\omega\, t) \quad \text{or} \quad v = V_m \sin(2\,\pi f\, t) \quad \text{where} \quad \omega = 2\,\pi f$$

$$f = 1/t \quad \text{or} \quad t = 1/f$$

Average value of a sine wave

$$V_{ave} = 0.636 \times V_m$$

RMS value of a sine wave

$$V_{rms} = 0.707 \times V_m$$

Peak–peak value of a sine wave

$$V_{peak\text{-}peak} = 2 \times V_m$$

Answers to test your knowledge questions

Chapter 4

4.1   1   (a) 120 m^2
            (b) 1.2×10^6 cm^2
            (c) 0.12×10^9 mm^2
    2   (a) 4000 cm^3
            (b) 0.004 m^3
            (c) 4×10^6 mm

4.2   1   7000 kg/m^3
    2   50 l
    3   9000 kg/m^3

4.3   1   490 J
    2   (a) 2.5 J
            (b) 2.1 J
    3   14.72 kJ

4.4   1   420 J
    2   4 m

4.5   1   (a) 80 N
            (b) 200 W
    2   (a) 450 W
            (b) 600 W
    3   (a) 1500 W
            (b) 510 mm/s

4.6   1   8.31 m
    2   281.25 kJ
    3   176.6 kJ, 176.6 kJ

4.7   1   80 mW
    2   (a) 500 V
            (b) 25 kW
    3   2 kW, 12 kWh, 84p
    4   £10.29

4.8	1	(a) 210 K
		(b) −48°C
	2	130 J/(kg °C)
	3	220°C

4.9	1	18.08 MJ
	2	24.70 MJ

4.10	1	16 A
	2	4 A

4.11	1	20 kΩ, 125 mA
	2	(a) 2400 Ω or 2.4 kΩ
		(b) 600000 Ω or 600 kΩ or 0.6 MΩ

4.12	1	1.44 Ω
	2	0.017 μΩm
	3	0.180 Ω

4.13		0.5 A, 4.5 V, 2.75 W

4.14	1	(a) 2 Ω
		(b) 4 A
	2	(a) all four in parallel
		(b) two in series, in parallel with another two in series
		(c) three in parallel in series with one
		(d) two in parallel, in series with two in series
	3	(a) 4 A
		(b) 44 A

4.15		1.818 A, 4.545 A, 2.728 A

4.16	1	$v = 50 \sin(628\, t)$
	2	10 ms
	3	29.4 V

4.17	1	35.35 mA
	2	311.1 V
	3	2.22 A

4.18	1	25 W
	2	0.632 A

4.19		(a) 46 min 36 s
		(b) 42.86 km/h
		(c) 45 km
		(d) 53.4 km/h

4.20	1	9.26×10^{-4} m/s^2
	2	1.42 s
	3	39.2 km/h

4.21   1   (a) $-5/9$ m/s^2
           (b) 750 N
       2   0.560 m/s^2

4.22   12.5 kg

4.23   90 rad/s, 16.2 m/s

4.24   (a) 0.4π rad/s^2 or 1.257 rad/s^2
           (b) 0.1π m/s^2 or 0.314 m/s^2
           (c) 187.5 revolutions

Chapter 5

5.1   1   (a) 3^7
           (b) 4^9
       2   (a) 2
           (b) 3^5
       3   (a) 7^6
           (b) 3^6
       4   (a) 9
           (b) ± 3
           (c) $1/2$
           (d) $\pm 2/3$

5.2   1   (a) 7.39×10
           (b) 1.1284×10^3
           (c) 1.9762×10^2
       2   (a) 2.401×10^{-1}
           (b) 1.74×10^{-2}
           (c) 9.23×10^{-3}
       3   (a) 1.7231×10^3
           (b) 3.129×10^{-3}
       4   (a) 2×10^2
           (b) 1.5×10^{-3}

5.3   1 $2\frac{1}{4}$
       2   $(a + b^2)/b$
       3   $a^2 c$

5.4   1 $5x - 7y$
       2   $2a^2 + ab - b^2$

5.5   1 $1 - a$

       2   $\dfrac{2}{3b} - 2b + 8$ or $2\left(\dfrac{1}{3b} - b + 4\right)$

       3   $S\,t$

       4   $p + 8p^2$ or $p(1 + 8p)$

5.6	1	4
	2	3
	3	−3
	4	9
	5	±4

5.7	1	$x = 1, y = 1$
	2	$p = -2, q = -3$
	3	$x = 0.3, y = 0.4$
	4	$x = 2, y = -3$

5.8	1	85.94 m^2
	2	$1.49 \times 10^{-11} \text{ N}$
	3	358.8 cm^3
	4	12.77 A

5.9   1 $m = F/a$

2   $\alpha = \dfrac{l_2 - l_1}{l_1 \theta}$

3   $g = \dfrac{4\pi^2 l}{t^2}$

4   $R = \dfrac{E - e - Ir}{I}$

5   $a = \sqrt{\left(\dfrac{rp}{x + y}\right)}$

5.10	1	$1\frac{1}{2}, -3$
	2	$-1/3, 4$
	3	$25.51, 1.490$

5.11	1	6.61 cm by 3.55 cm
	2	1.835 m or 18.165 m
	3	13.82 cm

5.12		(a) 0.4351
		(b) −0.8530
		(c) −3.4764

| 5.13 | 1 | 48 m |
| | 2 | 60.85 m |

5.14	1	$212° 57'$ and $327° 3'$
	2	$46° 50'$ and $313° 10'$
	3	$37° 17'$ and $217° 17'$

5.15   1 $B = 77°$, b = 18.16 mm, $c = 17.73$ mm, area $= 82.90 \text{ mm}^2$

2   $e = 32.91$ mm, $F = 43° 4'$, $D = 72° 56'$, area $= 393.2 \text{ mm}^2$

3   $X = 78° 8'$, $Y = 47° 13'$, $Z = 56° 38'$, area $= 44.04 \text{ cm}^2$

5.16   1   4.56 m, 5.38 m
    2   32.48 A, 14° 19'
    3   18.23 km/h

5.17   1   Gradient = ½
    2   (a) 7, −3
        (b) −2, 2/3
        (c) 4, 11
        (d) 3/2, −3/5
        (e) −2/9, −1/9
    3   6

5.18   (a) 40°C
        (b) 128 Ω
        (c) 139.27 Ω

5.19   –

Answers to review questions

Chapter 4

1   4500 kg/m^3

2   35 kg

3   50 kJ

4   2 kW

5   3 kW

6   –

7   –

8   18750 kJ

9   19620 kJ

10   10 W

11   28.8 kJ

12   –

13   502.5 kJ

14   –

15   12.5 V

16   125 W

17   (a) all four in series
      (b) all four in parallel
      (c) three in parallel connected in series with the
          remaining resistor
      (d) two in parallel connected in series with the two
          remaining resistors.

18   –

19    $v = 155.54 \sin(377t)$

20   (a) 150.65 V
      (b) −91.41 V

21   394 m/s

22   4.189 rad/s^2

Chapter 5

1   (a) 17
      (b) 3
      (c) 9
      (d) 4
      (e) 2.828
      (f) 6.240
      (g) 0.333

2   (a) 7.29×10^2
      (b) 1.5625×10^{-2}

3   10^8

4   (a) 2.173×10^3
      (b) 1.24×10^{-2}
      (c) 4.71×10^{-5}
      (d) 3.925×10^5

5   5×10^3

6   (a) ab
      (b) p^2q^3r

7    $5x^2 - xy^2 - 10x + 2y^2$

8   (a) $pq(1 + 2q)$
     (b) $3a^2(1 - 5b)$
     (c) $2ab(a - 2b + 3)$

9    $x = 10$

10   $x = -4$ or $x = 2$

11   $x = \pm 7/3$

12   $x = 1.7015$ or $x = -4.7015$

13   1.2 m

14   $\sin \theta = 5/13$, $\cos \theta = 12/13$, $\tan \theta = 5/12$

15   $\theta = 22.62°$

16   (a) 0.9272
     (b) −0.9877
     (c) −0.1763

17   –

18   $Y = 10.582$, $x = 19.579°$, $z = 32.421°$

19   12.766 sq cm

20   (a) 82.5°C
     (b) 26.1 m^3

21   –

Index